C0-AJW-968

# ROBOT TACTILE SENSING

# ROBOT TACTILE SENSING

*R. Andrew Russell*

*Department of Electrical and Computer Systems Engineering*

*Monash University*
*Victoria, Australia*

PRENTICE HALL

New York London Toronto Sydney Tokyo Singapore

1 2 3 4 5 94 93 92 91 90

Typeset by Keyboard Wizards, Professional Typesetters, Harbord, NSW.

Cover design by Norman Baptista.

Printed and bound in Australia by Impact Printing Pty Ltd,
Brunswick, Vic.

0 13 781592 1

National Library of Australia
Cataloguing-in-Publication Data

Russell, Andrew.
Robot tactile sensing.

Includes index.
ISBN 0 13 781592 1.

1. Robots. 2. Pattern recognition systems. 3. Tactile sensors.
I. Title.

629.892

Library of Congress
Cataloguing-in-Publication Data

Russell, Andrew (R. Andrew)
Robot tactile sensing / Andrew Russell.
p. cm.
Includes bibliographical references and index.
ISBN: 0-13-781592-1

1. Robots, Industrial. 2. Tactile sensors--Industrial applications.
3. Transducers. I. Title.

TS191.8.R87 1990 90-43830
670.42'72--dc20 CIP

Prentice Hall, Inc., *Englewood Cliffs, New Jersey*
Prentice Hall Canada, Inc., *Toronto*
Prentice Hall Hispanoamericana, SA, *Mexico*
Prentice Hall of India Private Ltd, *New Delhi*
Prentice Hall International, Inc., *London*
Prentice Hall of Japan, Inc., *Tokyo*
Prentice Hall of Southeast Asia Pty Ltd, *Singapore*
Editora Prentice Hall do Brasil Ltda, *Rio de Janeiro*

PRENTICE HALL

A division of Simon & Schuster

# Contents

# Preface

The human sense of touch provides us with an important source of information about our surroundings. Because of its unique position at the interface between our bodies and the outside world, touch sensing supplies sensory data which helps us manipulate and recognize objects and warn of harmful situations. Tactile sensing has the potential to fill a similar sensing role for robotic systems.

Developments in robot tactile sensing have been delayed by a relatively low level of research effort in this area. However, in recent years there has been a growing awareness of the importance of touch sensing in robotics. One of the factors limiting the spread of knowledge in this area has been the lack of appropriate courses which, in turn, is due partly to the absence of a suitable text book. This book was designed to fill the need for a student text to support courses in robot tactile sensing and to provide background material on tactile sensing for those involved in advanced, sensor-based robotics.

The material in the book is organized in the following manner. Human touch sensing represents a reference point, related to our own experience, for the study of robotic touch. Chapter 2 provides an overview of human touch sensing, dealing mainly with the types and location of sensor systems present in the human body. To give a conceptual framework for the chapters which deal with robot sensor designs, Chapter 3 introduces basic ideas related to sensors and transducers, and in particular the problems of addressing large sensor arrays. Chapters 4 through 7 describe the transducer mechanisms and construction of robot tactile sensors. This information has been grouped into separate chapters as follows:

Chapter 4 — measurement of the position of and forces applied by robot limbs;

Chapter 5 — contact sensor arrays;

Chapter 6 — compliant tactile sensors; and

Chapter 7 — other modes of tactile sensing.

It may be considered that the main function of robot hands and grippers is to grasp objects. However, they have an important role to play as providers of mobility for tactile sensors. From this perspective Chapter 8 presents an introduction to the design of robot hands and grippers.

The remaining Chapters 9 through 12 explain how tactile sensory information may be used to assist object manipulation and recognition operations. This material has been organized in the following manner:

Chapter 9 — control of robot motion by employing tactile feedback;

Chapter 10 — pattern recognition applied to tactile data;

Chapter 11 — active acquisition of touch sensory information; and

Chapter 12 — strategies for merging data from multiple sensors.

The tactile sensors described in this book will have important applications in the areas of medical rehabilitation, prosthetics, tele-operation, and tele-existence. However, these applications are outside the scope of this text.

This book is intended to support a senior-graduate-level course in robot tactile sensing. It would also be appropriate to a more general course in robot sensing covering computer vision, speech recognition, artificial sense of smell, etc., in addition to tactile sensing. The book will be of special interest to those who are concerned with applications of new sensory techniques and, in particular, to researchers in robotics and medical rehabilitation who are developing tactile sensor systems. I would like to thank Professor Ray Jarvis for his helpful discussions concerning the contents of Chapter 12, and also the reviewers and Kemal Ajay for offering corrections and improvements for this book.

My greatest thanks go to my wife, Glenyce, for her assistance, encouragement and support during the preparation of this book.

*R. Andrew Russell*
*Monash University, Australia*
*January 1990*

# 1 Introduction

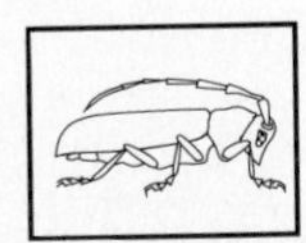
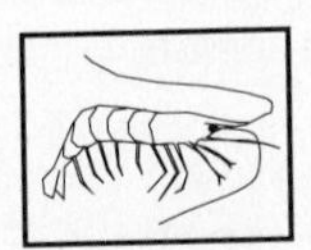

## 1.1 Why do industrial robots need sensory feedback?

The majority of robots presently employed in industry have no sensory capabilities. Their sequence of movements is prerecorded and this sequence is replayed, without variation, every time the robot performs a task. This lack of adaptability allows a robot to produce a consistent result, but only if the objects it is dealing with are equally consistent. A surprising range of tasks can be performed by following a prerecorded sequence of movements, including:

- spray painting;
- arc welding;
- palletizing;
- machine tending;
- spot welding;
- glue application;
- cutting; and
- assembly.

However, many potential robot applications require some form of sensing together with computer control which can modify the robot's actions based on the sensory information (Engelberger 1980: Nof 1985). The sensory information could be used to:

- locate randomly placed objects — reducing the need for expensive and inflexible jigs and fixtures;
- allow for variations in the shape of objects — an object only needs to be made as accurately as required for its proposed use, not as required for its automatic assembly. It would not be worthwhile building a car body accurate to 0.1 mm so that a robot could bolt things onto it by dead reckoning. Foodstuffs such as oranges or eggs will always vary in shape and size;

- protect against unexpected and dangerous situations, especially if the robot must work close to humans — an automatic guided vehicle (AGV) must not run into someone standing in its path;
- allow 'intelligent' recovery from error conditions — if a nut will not engage a thread try another nut before rejecting the partly completed assembly; and
- perform quality control — a robot's sensors can monitor the quality of the items it is manipulating. For example, a robot could examine an assembly for broken or missing parts.

Leaving aside hearing, taste, and smell the remaining forms of sensing can be divided into two groups — contact and non-contact sensors.

## 1.2 Comparing contact and non-contact sensors

Non-contact sensors measure interaction between the robot environment and:

- sound waves;
- light and other electromagnetic waves; and
- electric, magnetic, and electrostatic fields.

Perhaps because vision so dominates our human senses, the majority of work in the area of robot sensing has concentrated on computer vision. Vision systems are the only sophisticated forms of sensing which have found relatively wide application in industry.

However, many creatures including humans make good use of the tactile information they obtain through physical contact with external objects. Tactile sensing is very direct: it is not distorted by perspective, confused by external lighting or greatly affected by the material constitution or surface finish of objects. We humans use tactile information to maintain the posture of our bodies, to provide a warning of physical danger, and to monitor walking and grasping. Creatures such as the invertebrates and nocturnal animals have highly developed tactile senses to detect their prey and avoid danger. Perhaps our primitive robots could likewise benefit from a sense of touch.

Harmon defines tactile sensing as 'the continuous-variable sensing of forces in an array' (Harmon 1982). This is a restricted definition which ignores many other forms of sensing mediated by touch. Here is a more general definition which defines the scope of this book:

> **Tactile sensing covers any sensing modality which requires *physical contact* between sensor and an external object.**

## 1.3 Applications for robot tactile sensors

Most robot manipulation and assembly tasks would benefit from the utilization of tactile sensory information. When lifting an object, tactile sensing could detect the onset of slip in time for corrective action to be taken. When moving objects in close proximity the difference between a clearance of +1 mm and –1 mm is difficult to determine visually. However, this difference is immediately obvious if the resulting forces are monitored. In confined spaces, where vision is impractical, touch can be used to locate and recognize objects. Consider finding a particular denomination of coin in a pocket containing assorted coins, a bunch of keys, a comb, and a pocket handkerchief. For recognition tasks several object properties are readily determined by touch including:

- weight;
- temperature;
- thermal characteristics of the material making up the object;
- compliance, resilience and coefficient of friction of the object surface;
- electrical conductivity; and
- surface texture.

Other properties such as location, orientation, size, and surface shape can be determined either visually or by touch. Tactile sensing also provides information about forces and torques transferred between the robot hand or gripper and objects it is manipulating.

## 1.4 What does tactile sensing involve?

Tactile sensing comprises a complex web of interrelated functions, both in robots and living creatures. For this reason it will be necessary to consider the interaction of several related topics. As you would expect we will cover the design of 'skin-like' touch sensors. However, a touch sensor by itself is almost useless. In order to gather information it also requires:

- some means of locomotion to bring the sensor into contact with the object being explored;
- additional sensors to determine the position and orientation of the touch sensor; and
- some form of 'intelligence' to plan the search for tactile information and to interpret the results.

Therefore, we will consider grippers, actuators, transducers, and information processing, as well as the sense of touch in humans, animals, and machines.

The state-of-the-art in touch sensing technology for robots, tele-operation, and prosthetics falls well behind the popular image portrayed in science fiction films. Present capability in these areas is very primitive. There is a need for much fundamental research to develop improved transducers, analyze their output and use the resulting information in object recognition and manipulation operations.

Eventually robots will 'break away' from their well-organized corner of a factory and start dealing with the relatively unstructured environment in the rest of the world. When this happens they will need a whole range of sensory information, including tactile information, to navigate, to manipulate objects, to protect themselves, and to protect their surroundings.

## Bibliography

DARIO, P., and DE ROSSI, D., 'Tactile Sensors and the Gripping Challenge', *IEEE Spectrum*, Vol. 22, No. 8, August 1985, pp. 46–52.

ENGELBERGER, J. F., *Robotics in Practice*, Kogan Page, London, 1980.

HARMON, L. D., 'Automated Tactile Sensing', *The International Journal of Robotics Research*, Vol. 1, No. 2, Summer 1982, pp. 3–32.

HARMON, L. D., 'Tactile Sensing for Robots', in *Recent Advances in Robotics* (Beni and Hackwood, eds), John Wiley & Sons, New York, 1985, pp. 389–424.

NOF, S. Y., ed., *Handbook of Industrial Robotics*, John Wiley & Sons, New York, 1985.

## Questions

1.1 Make a list of industrial robot applications which require little or no sensory feedback.

1.2 For each application you listed in answer to Question 1.1, describe how you think the need for sensory feedback can be avoided.

1.3 Make up a list of tasks that could only be performed by an industrial robot system incorporating sensory feedback.

1.4 For each of the tasks you proposed in answer to Question 1.3, explain why sensory feedback would be essential for a robot to perform the task.

1.5 What physical properties can be readily measured by tactile sensors?

1.6 Describe how you think tactile sensing is involved in:

(a) throwing a ball;
(b) walking; and
(c) finding a light switch in the dark.

1.7 What sensory capabilities do you think an autonomous domestic carpet-cleaning robot would require? Provide a justification for each of the sensors you choose.

1.8 If you lost the sense of touch in your arms and hands, which tasks do you think you would no longer be able to perform?

1.9 You are given an object which you do not recognize and which is small enough to be easily handled. Apart from looking at it from all angles, what tests do you apply to the object to find out its material constitution and what it might be used for?

# 2 Human touch sensing

The human hand is remarkably good at grasping things, manipulating them, and determining many of their physical properties. All of these capabilities are made possible by information from the senses (see Figure 2.1), especially the sense of touch.

Sensory information about the position and nature of contact with a grasped object is provided by sensors in the skin. The position of the fingers and the forces they apply are measured by nerve endings in the muscles, tendons, and surrounding tissues. At the lowest level, the spinal cord uses sensory information to perform purely reflex control of the hand. One of these reflexes counteracts involuntary stretching of the muscles and thus helps to

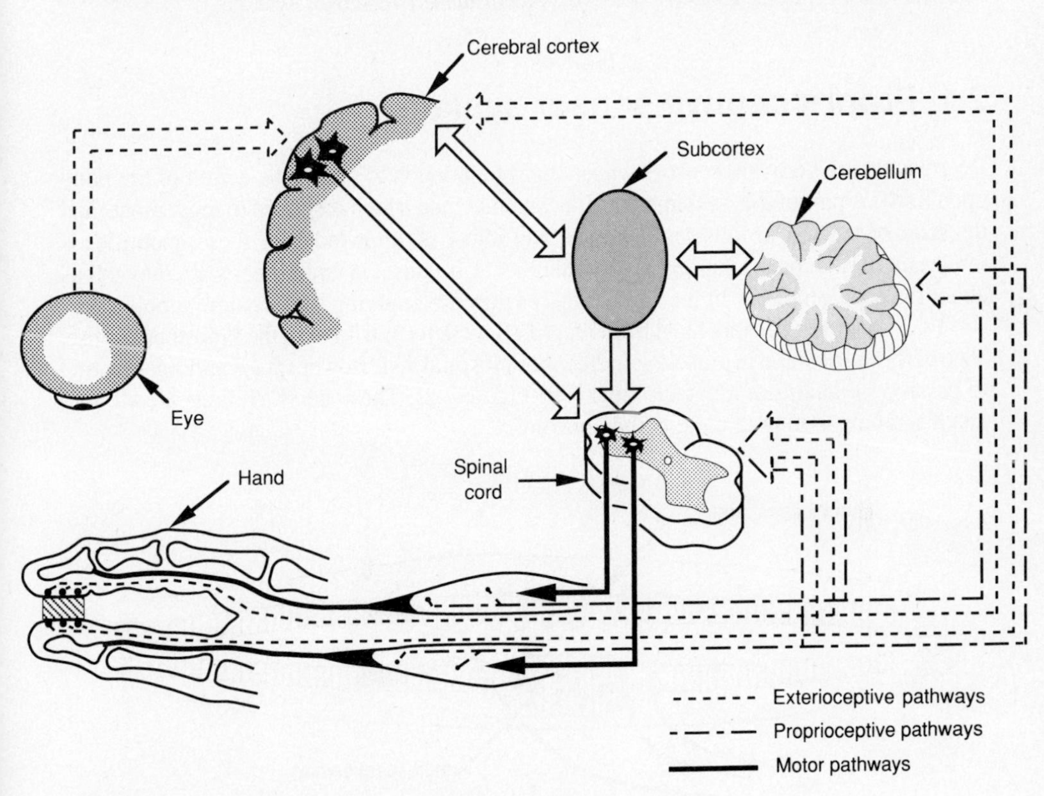


**Figure 2.1** The sensory control of grasping and manipulation
(Tubiana 1981, redrawn with permission of the publisher)

maintain the posture of the hand. Higher levels in the nervous system, the cerebellum, subcortex, and cerebral cortex, exercise progressively more sophisticated control. At the cerebral cortex level, lengthy tasks with complex goals can be organized, perhaps using visual inspection to augment the touch senses (Albus 1979).

In robotic touch sensing we are trying to emulate some of the capabilities of the human touch sensory system. For this reason it is informative to find out about the sensing modalities involved in human touch sensing and their capabilities. We are unlikely, at least in the short term, to be able to reproduce the detailed structure of biological sensors. However, the existence of these biological systems and our own experience of using them, show us what developments are possible and how they may be applied.

The human body has two sensory systems which react to contact with external objects. These are:

1. the proprioceptive system which measures internal quantities such as limb joint angles, muscle extension, muscle tension, etc. — these quantities are indirectly related to contact with external objects; and
2. the exterioceptive system which reacts to the changes in temperature and deformation of the skin surface — these quantities are a direct result of contact with external objects.

The rest of this chapter gives a brief overview of these two sensor systems.

## 2.1 Proprioceptive sensor system

Deep inside the body are sensors whose main function is to detect the action of the body upon itself. A part of this system is the kinesthetic sense which responds to movements and the state of tension in muscles, tendons, and joints. A knowledge of these quantities is important for the manipulation of external objects, because, in an indirect way, they tell us about the position of our limbs and the forces they are applying to the external objects.

Four groups of sensors have been identified, which contribute to the kinesthetic sense. Two of these are found in muscle spindles. Annulospiral and 'flower spray' endings respond to passive stretching of a muscle fiber (see Figure 2.2). These nerve endings are slow to adapt and help to maintain the body's posture.

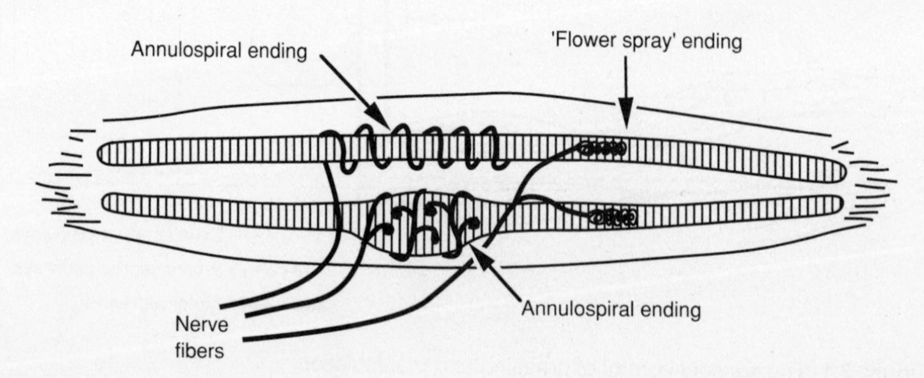


**Figure 2.2** Neuromuscular spindle

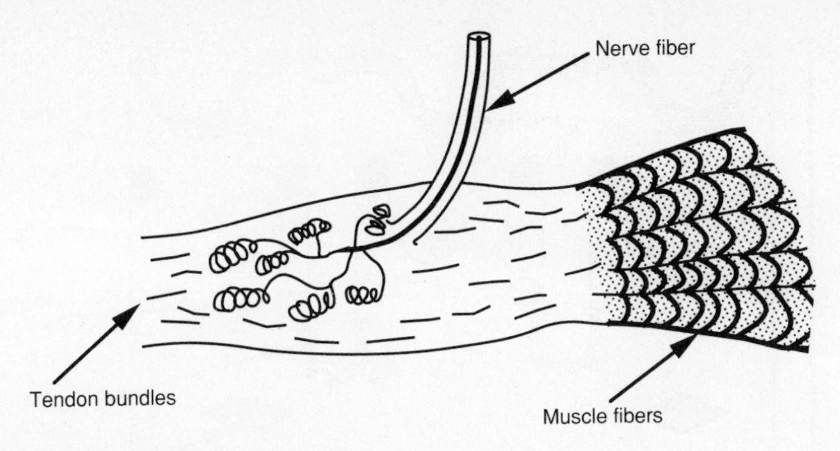


**Figure 2.3** Neurotendinous spindle

Total muscle tension is sensed by the third group of nerve endings which is located in muscle tendons (see Figure 2.3). This group responds to muscle tension produced either passively by an external force or actively by the muscle itself.

The fourth group of nerve endings is situated in the connective tissue surrounding the joints, and measures limb joint angles. This group's ability to detect small angular movements of the joints is remarkably good. For instance, displacements of about half a degree taking place in less than two seconds can be discerned at the shoulder joint.

## 2.2 Exterioceptive sensor system

The skin is made up of two main layers: the epidermis, which is the outermost protective layer incorporating the fingerprint papillary grooves of the palm and fingers, and the dermis which lies between the epidermis and deeper layers of subcutaneous tissues and fat layers. The nerve endings of the exterioceptive sensor system are mainly confined to dermal and epidermal layers and consist of a variety of specialized nerve endings. They provide a direct response to contact pressure, texture, temperature, and pain. Figure 2.4 shows a schematic representation of the nerve supply to the outer layers of the hair-free skin covering the fingers and palm.

Four different kinds of sensors register deformation of the skin caused by contact with external objects. Meissner and Merkel endings lie close to the skin surface and have high spatial resolution. In the fingerprint skin of the hand Meissner endings are located between the papillary ridges of the dermis, while the Merkel endings are located at the ends of these ridges. Pacinian corpuscles and Ruffini endings are embedded deep in the skin and hence their receptive fields are much broader. In addition, there are free nerve endings with a high threshold which respond to painful and potentially harmful stimuli, such as pricking with a sharp needle. Table 2.1 shows more of the characteristics of the four different types of sensor.

All nerve endings, except those signaling pain, exhibit adaptation where the strength of a steady stimulus diminishes with time. The precise rate of adaptation affects the frequency

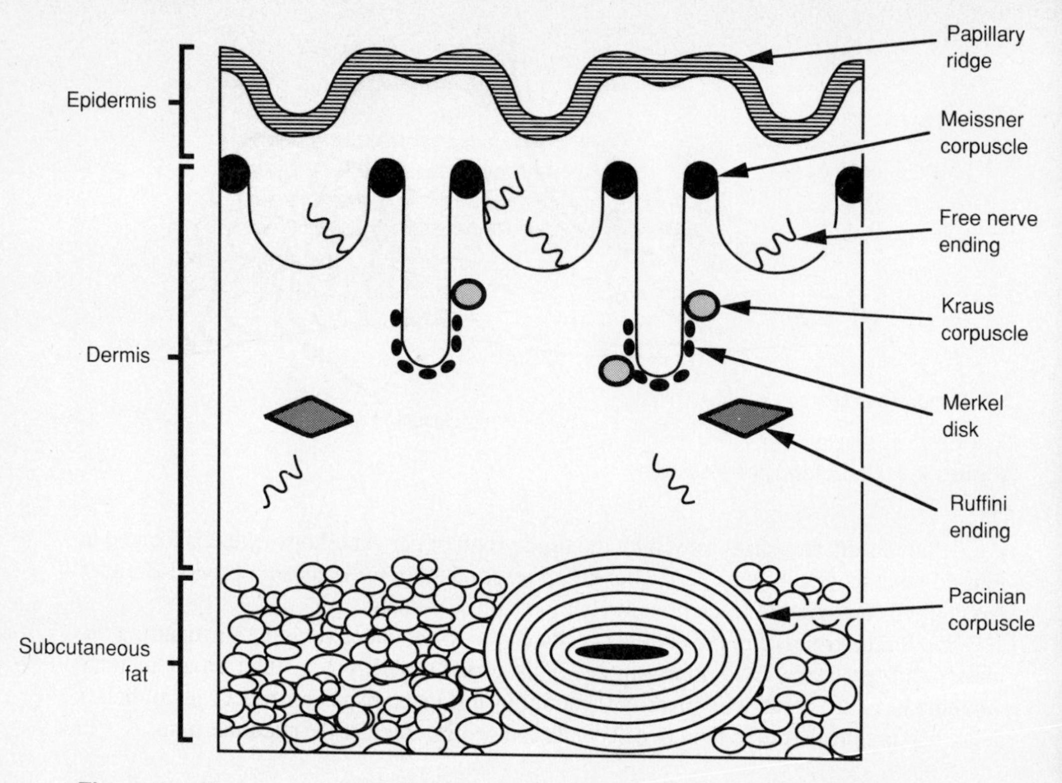


**Figure 2.4** Nerve supply to the fingerprint skin of the hand

**Table 2.1** The characteristics of some nerve endings which respond to small deformations of the skin below the threshold of pain

| *Nerve ending* | *Receptive field (mm)* | *Vibration response (Hz)* | *Stimulus* |
|---|---|---|---|
| Meissner | 3 – 4 | 8 – 64 | Texture, normal force |
| Merkel | 3 – 4 | 2 – 32 | Shape, edges, texture |
| Pacinian | 10 | 64 – 400 | Vibration |
| Ruffini | — | 1 – 16 | Lateral skin stretch |

response of each type of receptor. After swimming in cold water for a few minutes, the water does not feel as cold as it did at the start. The nerve endings signaling temperature have adapted. Temperature is relayed by two groups of sensors, one responding to 'cold', the other to 'warmth' (see Table 2.2).

The type of stimulus which activates a particular nerve fiber is a property of the membrane surrounding the nerve ending. Details of the form of the nerve ending — bare nerves as in the free nerve endings; endings with expanded tips, such as Merkel's disks; or

**Table 2.2** The characteristics of some nerve endings which respond to temperature

| *Nerve ending* | *Temperature range (°C)* | *Stimulus* |
|---|---|---|
| Kraus | 18 – 40 | Cold |
| Ruffini | 40 + | Warm (normal body temp. 37°C) |

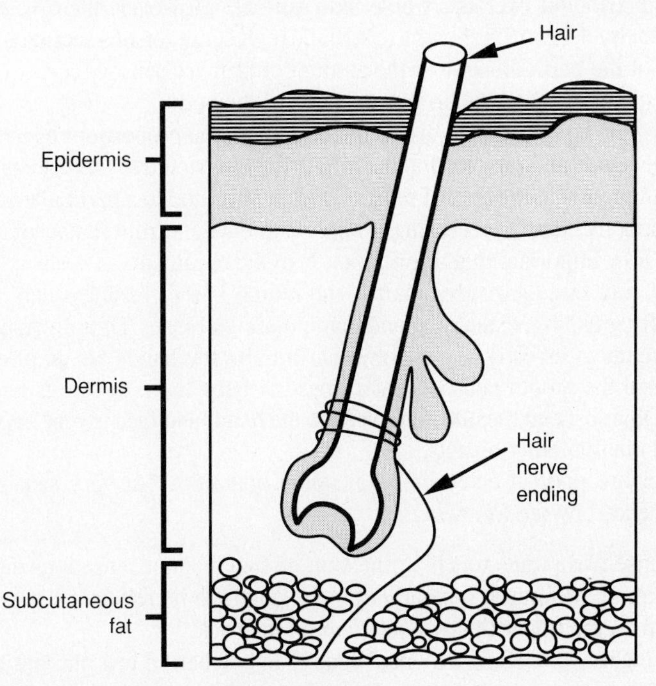


**Figure 2.5** Nerve supply to the hair follicle

encapsulated endings, such as Pacinian corpuscles — influence the dynamic range of mechanical stimuli to which the receptor responds. Another factor which modifies the response of mechanoreceptors is the manner in which they are attached to the surrounding tissue structures.

## 2.3 Body hairs as sensors

Over the majority of the skin surface area body hairs play the role of short-range proximity sensors in addition to their function of retaining body heat. Hairs are exquisitely sensitive to touch, acting as levers in which slight changes in position are detected and signaled by a network of nerve endings surrounding the hair follicle (see Figure 2.5). Those hairs which are not usually covered by clothing are particularly sensitive to the slightest touch or a gentle air flow.

Although the skin is the most accessible organ there are still many details about the sensing functions of specific nerve endings which are unknown. Some nerve endings respond to more than one stimulus. This makes it difficult to be sure which nerve endings contribute to a particular sensation.

## 2.4 The distribution of tactile sensors

Tactile sensors are distributed over the whole skin surface. However, there is a great variation in sensor density. Figure 2.6 shows a schematic representation of a section through the left parietal lobe of the brain illustrating the amount of primary sensory cortex (length of dark lines) that responds to pressure on various parts of the body.

The caricatures of the body parts are drawn in about the same proportions as the lines. You will note that the hands and face around the mouth are connected to large areas of the sensory cortex. It is thought that the area of sensory cortex allocated to a particular sensory input indicates how much cerebral processing is applied to the data from that source. This in turn may indicate how important that information is to the organism.

In most animals, and insects for that matter, the mouth is the 'business end' of the creature and is used for grasping, examining and manipulating objects. Thus a young baby starts by using its mouth to investigate new objects. Initially the hands are employed to transport the objects to the mouth and only later does the baby learn to use its hands to explore new things. Obviously tactile information from the hands and face is very important for investigating and manipulating objects.

The touch senses are poor at determining absolute quantities but very sensitive to changes. This is evidenced by the following:

- The ability to sense two points touching the skin as two distinct sensations depends upon their separation. Minimum separation for the points to be felt as distinct varies from about 1 mm on the finger tip to about 2 cm on the back.
- By running the finger over a surface a 'step' change in height of a few microns can be detected. Try folding over the corner of a sheet of paper. The resulting step change of one thickness of paper can be readily detected. A sheet of 80 gsm copying paper has a thickness of about 0.1 mm.
- Place your left hand in water at 4°C and your right hand in water at 40°C. When both hands are then placed in water at 20°C the left hand feels warm and the right feels cold.
- The skin of the face can detect changes of temperature of the order of 0.01°C per second.

Individual or groups of nerve endings channel into one nerve fiber which does not synapse (connect to other nerves) until it reaches the spinal cord. Therefore there is no possibility of neural processing of tactile data close to the skin. This is contrasted with the eye where lateral inhibition of optical nerve fibers in the eye performs edge enhancement and other processing tasks. However, there is evidence that the physical structure of the skin and the position of the nerve attachments perform edge enhancement of information about skin indentation. Little is known about the processing of tactile information at higher levels in the nervous system.

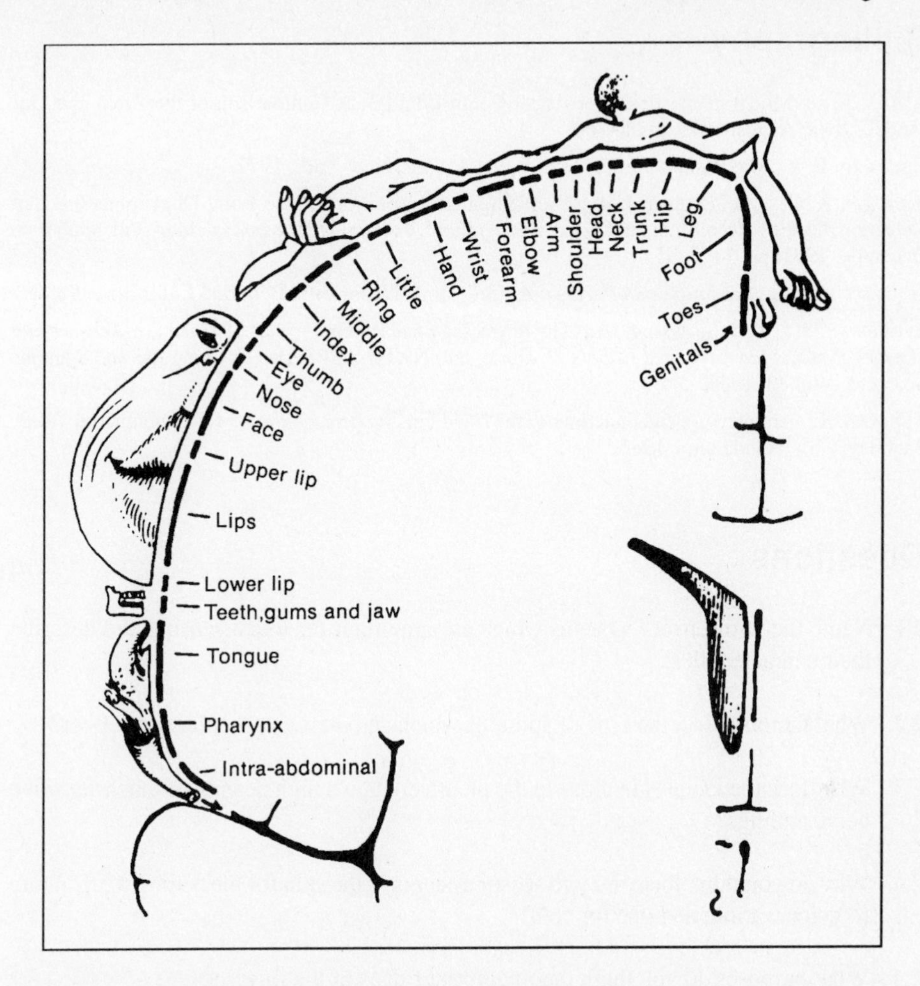


**Figure 2.6** Primary sensory cortex processing data from different parts of the body

(Reprinted by permission of Macmillan Publishing Company from *The Cerebral Cortex of Man* by Wilder Penfield and Theodore Rasmussen. Copyright 1950 by Macmillan Publishing Company; copyright renewed © 1978 by Theodore Rasmussen.)

## 2.5 Conclusion

There is much about the human touch sensory system which is still unknown. However, we have a broad idea of the sensing techniques which are being employed and the capabilities of the sensors. Therefore, it is worthwhile to attempt to develop sensors which parallel the human tactile sensors and investigate techniques for using their output to aid recognition and manipulation operations.

## Bibliography

ALBUS, J., 'A Model of the Brain for Robot Control. Part 3: A Comparison of the Brain and Our Model', *Byte*, August 1979, pp. 66–80.

GELDARD, F. A., *The Human Senses*, John Wiley & Sons, New York, 1972.

JOHNSON, K. O., and PHILLIPS, J. R., 'Tactile Spatial Resolution I. Two-Point Discrimination, Gap Detection, Grating Resolution, and Letter Recognition', *Journal of Neurophysiology*, Vol. 46, No. 6, December 1981, pp. 1177–91.

GOLDSTEIN, E. B., *Sensation and Perception*, (3rd edn.), Wadsworth, Belmont, California, 1989.

LEDERMAN, S. J., and BROWSE, R. A., 'The Physiology and Psychophysics of Touch', in *Sensors and Sensory Systems for Advanced Robots*, P. Dario, ed., NATO ASI Series F: Computer and Systems Sciences, Vol. 43, 1988.

TUBIANA, R., 'Architecture and Functions of the Hand', in *The Hand,* Volume 1, R. Tubiana, ed., W.B. Saunders Co., Philadelphia, 1981.

## Questions

2.1 Name the two sensory systems which are important for touch sensing and describe their characteristics.

2.2 What factors affect the type of stimulus which activates a particular nerve fiber?

2.3 Why does the facial skin close to the mouth contain a high density of touch sensitive nerve endings?

2.4 Why do you think there are two sensor systems in the skin for measuring temperature (one for warmth and one for cold)?

2.5 What purposes do you think the fingerprint ridges of the finger serve?

2.6 Describe sensing functions performed by the fingernail.

# 3 Sensors and transducers

This chapter introduces the concepts of carriers of information and the physical effects which transfer information from one carrier to another. The process of transferring information from one carrier to another is fundamental to the operation of all sensors and transducers. In many applications, tactile sensors will consist of large arrays of sensor elements. Methods are examined for addressing individual sensor elements in an array and transferring their output to processing circuits either within the sensor itself or externally.

## 3.1 Information-processing systems

An information-processing system can be thought of as comprising three linked units as shown in Figure 3.1.

Information is acquired by the input transducer, often called the sensor, and converted into a form compatible with the processor. After modification by the processor the resulting information is again converted into a form suitable for output to the outside world. Thus a sensor is an input transducer which converts information in some physical manifestation into a form which can be processed. A robot can be considered as an information-processing system where the input transducer could be a tactile sensor mounted on the gripper, the processor a computer controlling the robot, and the actuators providing the system output correspond to the robot servos. Such a system could modify the motion of the robot based on information received from the tactile sensor.

What is information? From the information theoretic point of view, first expounded by Claude Shannon, information is a quantity which reduces uncertainty. From a personal viewpoint we might think of information as something which increases our knowledge.

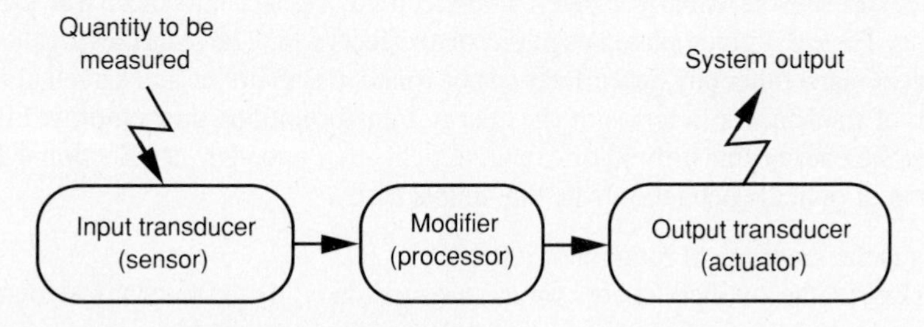


**Figure 3.1** An information-processing system

These two definitions give quite different, but not contradictory, views of information. However, one of the important characteristics of information is that it can only exist as either a form of energy or some material manifestation. This means that there can be no transfer of information without a corresponding transfer of energy or matter.

## 3.2 Carriers of information

We can identify six broad groupings of carriers of information:

1. *Radiant* — concerned with electromagnetic waves of all frequencies, ranging from radio waves to gamma-rays. The main parameters in which we are interested are intensity, frequency, polarization, and phase.
2. *Mechanical* — deals with all kinds of external parameters of matter such as position, velocity, size, thickness, force, and vacuum.
3. *Thermal* — involves temperature, temperature gradient, heat, and entropy.
4. *Electrical* — concerned with electrical parameters such as voltage, current, resistance, capacitance, etc.
5. *Magnetic* — includes field intensity, flux density, and permeability.
6. *Chemical* — deals with the internal structure of matter. Its main parameters are the concentration of a certain material, the crystal structure, and aggregation state.

In a transducer, a physical effect is used to translate between the signal domain of the input and the signal domain of the processor. Table 3.1 lists some solid state physical effects which take an input in either the radiant, mechanical, thermal, magnetic, or chemical domain and produce an electrical output.

There are several signal domains where information-processing could be performed:

- *Mechanical* — by means of fluidic components. (Speed of operation of fluidic components is limited by the velocity of sound in the working fluid, around $10^3$ m/s.)
- *Electrical* — using electronic circuits. (Speed of operation of semiconductor-based circuits is limited by the mobility of charge carriers in the semiconductor material, around $10^5$ m/s.)
- *Radiant* — by means of optical components. (Speed is limited by the speed of light in a light guide, around $10^8$ m/s.)

However, at the present time the electrical signal domain can be manipulated best. Therefore, the sensors usually convert non-electrical signals into electrical signals for processing. Table 3.1 gives examples of physical effects which have an electrical output. A collection of many other physical effects can be found in Ballantyne and Lovett (1980). We can think of transducers in terms of the energy transformations they employ. Figure 3.2 illustrates the energy transformations in an optical shaft encoder (see Section 4.1.2 for a description of optical encoders). Note that in this case:

- input is the mechanical rotation of a shaft;
- light beams, the auxiliary energy source, are modulated by the mechanical rotation; and
- the modulated light is converted into an electrical output signal.

**Table 3.1** Solid state physical effects with electrical output (From Middlehoek and Hoogerwerf 1986)

| *Domain* | *Measurand* | *Stationary* | *Gradient* | *Time derivative* |
|---|---|---|---|---|
| Radiant | Intensity<br>Frequency<br>Polarization<br>Phase | Photoelectric effect<br>Photoconductivity<br>Photodielectric effect | Lateral photoelectric effect | |
| Mechanical | Force | Piezoelectric effect<br>Piezoresistance | | Acoustoelectric effect |
| | Velocity<br>Size<br>Place | Place dependence of electrical field | | |
| Thermal | Temperature | Temperature dependence of all electrical properties of matter | Seebeck effect<br>Ettinghausen-Nernst effect | |
| Magnetic | Field intensity | Hall effect<br>Magnetoresistance<br>Superconductivity<br>Suhl effect<br>Ettinghausen-Nernst effect | | Faraday-Henry Law |
| Chemical | Concentration<br>Crystal structure<br>Aggregation condition | All electrical properties of matter | Volta effect | Redox reaction |

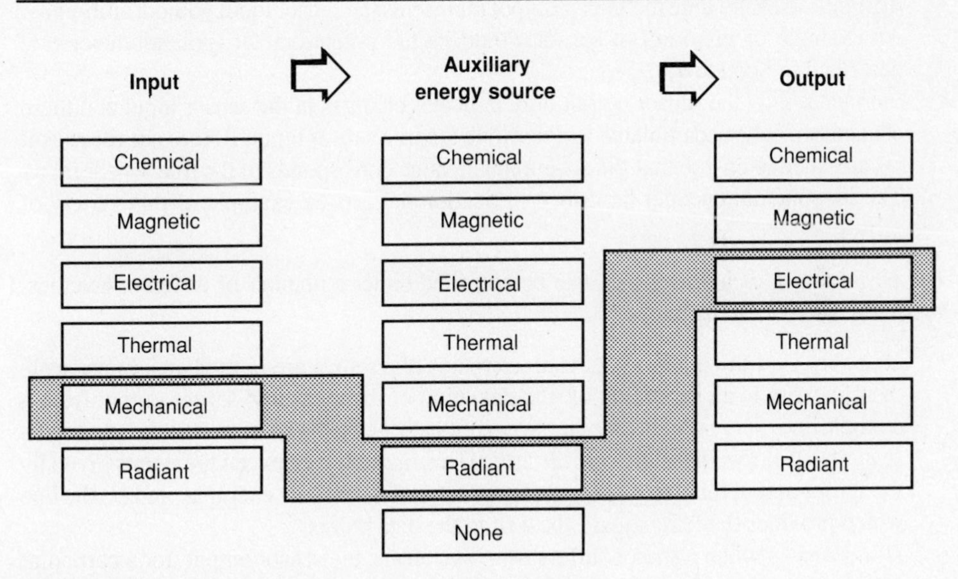


**Figure 3.2** Energy transformations within a transducer

In some cases, it may be necessary to measure the space or time derivative of a quantity. Instead of measuring the quantity directly and then calculating derivatives it may be possible to choose a transducer effect which produces the place or time derivative directly. For example, to determine the rotational velocity of a shaft, the time derivative of the output of a shaft encoder can be calculated. Alternatively, a DC tachometer generator directly produces an output proportional to rotational velocity. The tachogenerator is a DC electrical generator whose output voltage is proportional to the rotation speed of the armature.

## 3.3 Performance and characteristics

There are several categories which are useful for classifying different sensors. These include:

- *Self-generating* — energy from the input signal is directly converted to the output signal. Selenium photoelectric cells are an example of self-generating sensors. Some lightmeters contain a selenium photocell connected to a specially calibrated moving coil meter. This kind of sensor has the following advantages: (1) it has no offset (produces zero output for zero input); and (2) it requires no auxiliary power source (good for low power applications).
- *Modulating* — the main energy supply is not the input signal but an auxiliary energy source. In a strain gauge bridge, mechanical strain changes the electrical resistance of the bridge elements (see Section 4.3.1 for a description of strain gauges). An electrical current must be passed through the bridge elements to produce the output signal — a varying voltage. Modulating sensors are usually better at detecting weak signals.
- *Absolute* — at any time the sensor output represents the sensor input without ambiguity. An example of this kind of sensor would be the potentiometer (potentiometers are described in Section 4.1.1).
- *Incremental* — the sensor output only indicates changes in the sensor input and these changes must be accumulated to determine the true sensor input. Before use the sensor is calibrated to ensure that the accumulated value corresponds to the true sensor input. The incremental encoder described in Section 4.1.2 is an example of this variety of sensor.

The performance of a sensor can also be specified under a number of different headings which include:

- *Linearity* — if the input/output characteristics of a sensor are plotted on a linear scale then linearity is an indication of the deviation of the measured sensor output from a straight line. The straight line approximation to the sensor output can be chosen in several ways. Two possible choices are: (1) a straight line between the points given by the sensor output for zero input and sensor output for 100 per cent input; and (2) the line which provides the least squares best fit to the data points.
- *Hysteresis* — when a sensor suffers from hysteresis, the sensor output, for a particular value of sensor input, varies depending upon whether the sensor input was increasing or decreasing when the measurement was made.

- *Repeatability* — when the same input is applied to a sensor, repeatability is a measure of the variability in the sensor output. This is sometimes referred to as precision.
- *Resolution* — is the smallest change in sensor input which produces a detectable change in sensor output.
- *Sensitivity* — a small change in the sensor input will cause a corresponding small change in sensor output. Sensitivity is the change in output divided by the change in input.
- *Noise* — is the level of spurious signals occurring in the sensor output which are not caused by the sensor input.

Usually, sensor signals will be processed some distance away from the sensor. This may be due to space limitations close to the sensor or because the environment surrounding the sensor is unsuitable for sensitive electronic components. Heat, shock, and vibration are some factors which damage electronic components. The transmission of sensor signals, especially from large arrrays of sensor elements, needs special attention.

## 3.4 Sensor addressing

To produce a sensory skin we must form an array of sensors over the two-dimensional area of the skin surface. This presents the problem of how to read data from each individual sensor. In the human skin the nerve from each sensor or group of sensors does not synapse (pass its signal on to another nerve) until the spinal column. This indicates that there is one data channel per sensor. Such an arrangement works well for biological systems but is less suitable for human-made instruments. The manufacture of reliable, flexible, and compact wiring for a flexible artificial skin is a difficult technological problem. Routing a great number of wires through the articulated joints of a robot hand also poses physical problems. Therefore, there are many advantages to be gained by minimizing the number of wires to a sensor array.

Many capacitative, piezoelectric, and piezoresistive tactile transducers are two-terminal devices (see Figure 3.3). So to illustrate some addressing schemes it is assumed that the sensor array contains such two-terminal sensors.

In applications where electrical noise pick-up is a problem each sensor can be connected to external circuitry by a twisted pair of wires or coaxial cable. To provide screening from interference, the coaxial cable outer screen or one of the two twisted pair connections could be grounded. Improved noise rejection can be obtained by connecting the two sensor wires to a balanced amplifier. The obvious disadvantage of this addressing scheme is the large

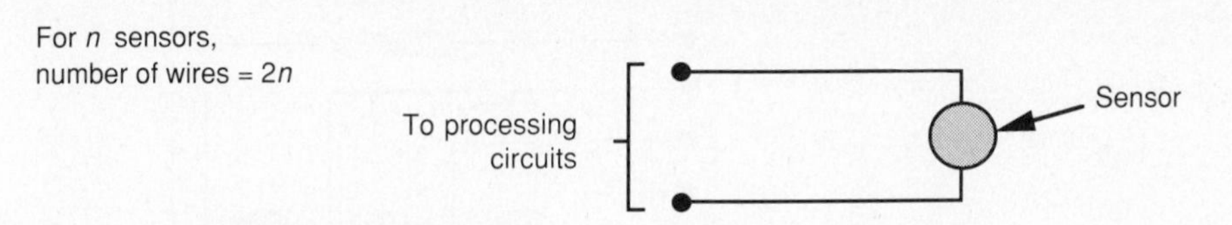


**Figure 3.3** Sensor addressing with two wires per sensor

number of connecting wires required. For example, an array of 100 sensors would require 200 wires.

The number of connections may be reduced by employing a common return wire to each sensor (see Figure 3.4). In this case an array of 100 sensors would require 101 wires. Piezoelectric sensors are often addressed using this technique because the more economic 'row and column matrix' shown in Figure 3.6 is unsuitable.

Potential divider addressing (Figure 3.5) can be used with piezoresistive sensors in applications where the wiring has appreciable resistance, especially where the wiring resistance may vary. (In this case an array of 100 sensors would require 103 wires provided it was feasible to use only one divider chain.) A known current $i$ is injected into the potential divider chain. To read the resistance $r_x$ of sensor element $x$ the potentials $v_1$ and $v_2$ are measured. The current through element $x$ is $i$ and the potential across $x$ will be $v_1 - v_2$ independent of the resistance of any of the wiring in the circuit. Therefore $r_x$ can be found:

$$r_x = (v_1 - v_2)/i \tag{3.1}$$

Row and column matrix addressing (Figure 3.6) gives a large reduction in the number of connecting wires. (An array of 100 sensors would require only 20 wires.) Unless transducers only conduct in one direction the effect of parasitic conduction paths distorts the data. If the effect of parasitic conduction paths can be ignored then:

$$V_{out} = (V.r_x)/(R + r_x) \tag{3.2}$$

Parasitic paths in row and column addressing can be counteracted by the use of special circuits.

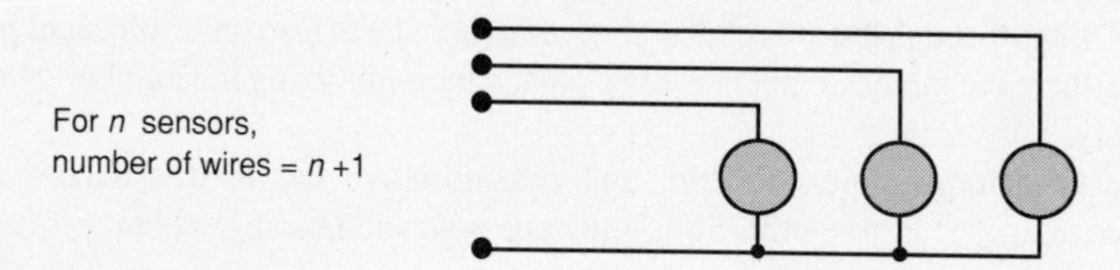


**Figure 3.4** Sensor addressing with one wire per sensor and a common return

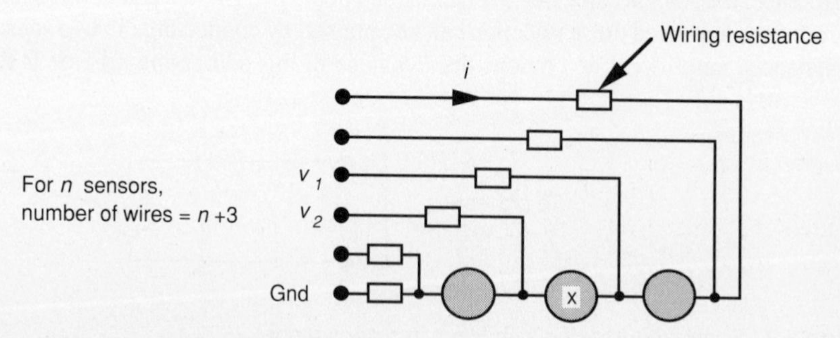


**Figure 3.5** Potential divider chain

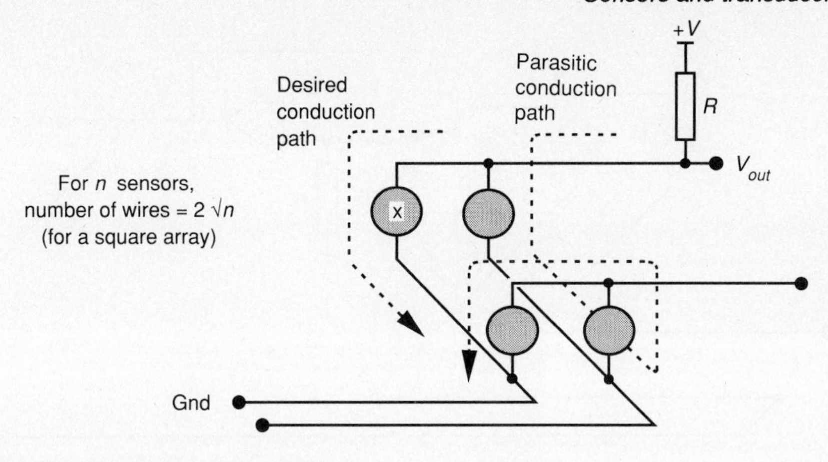


**Figure 3.6** Row and column matrix

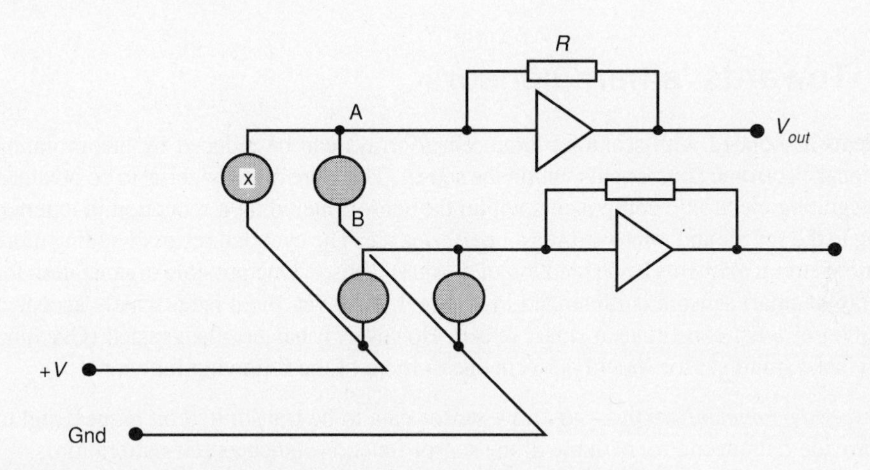


**Figure 3.7** Eliminating parasitic paths using a virtual earth

In Figure 3.7 negative feedback causes the input to the amplifier A to be maintained at near ground potential (a virtual earth). All column lines except the active one are held at ground potential B. Because points A and B are at the same potential no current can flow in the parasitic path. The resistance of sensor element $r_x$ is given by:

$$r_x = -(V.R)/V_{out} \tag{3.3}$$

An alternative technique for eliminating parasitic paths is illustrated in Figure 3.8. A unity gain amplifier takes the sensor output voltage A and feeds it back on all unused column lines. This makes the potential at A the same as at B and therefore no current flows in the parasitic conduction path. The value of $r_x$ is found using Equation 3.2.

Both of these methods require additional circuitry but the problem of parasitic paths is eliminated providing that the wiring to the sensor array has negligible resistance compared to the sensor elements.

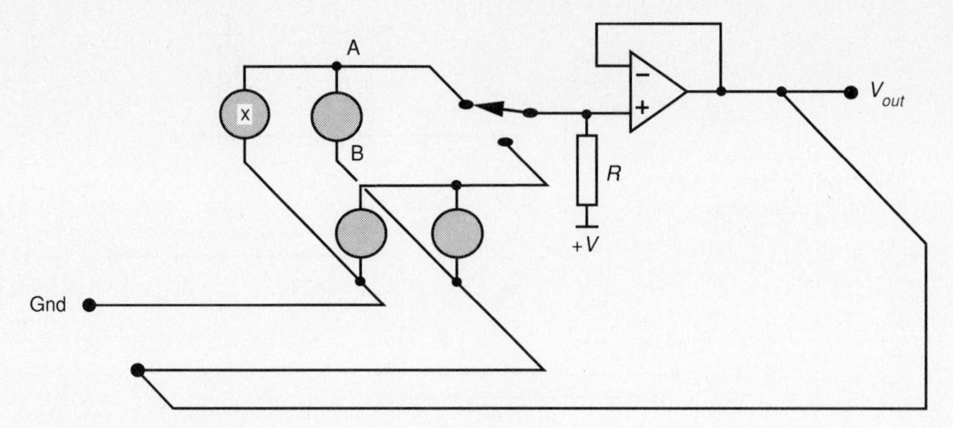


**Figure 3.8** Eliminating parasitic paths by guarding

## 3.5 Towards 'smart' sensors

Problems associated with scanning large sensor arrays can be reduced by incorporating additional electronic components within the sensor. There are other benefits to be obtained by integrating electronic components within the sensor, including a reduction in external wiring to the sensor and improved sensor performance. The eventual result of adding more and more circuit elements is the creation of a 'smart' sensor. One possible organization for an array of smart sensors is illustrated in Figure 3.9. As yet, there is no widely accepted definition of what constitutes a smart sensor. However, it has been suggested (Giachino 1986) that a smart sensor should perform one or more of the following functions:

- *two-way communication* — to allow sensor data to be transmitted on request and to provide remote control of some of the sensor functions (such as self-calibration);
- *self-calibration* — to compensate for the change of sensor characteristics over time;
- *computation* — to compensate for environmental changes (such as temperature), and to perform data processing and reduction so that a smaller quantity of information need be transmitted (for instance, only transmit significant changes in the sensor readings);
- *multisensing* — to measure more than one kind of physical or chemical quantity simultaneously. For example, an artificial skin sensor capable of measuring the temperature of an object as well as the pattern of pressure applied to the skin.

The path towards creating smart sensors involves the progressive incorporation of electronic circuits within the sensor structure. Adding a diode in series (see Figure 3.10) with each sensing element could be considered to be a first step in this direction (Snyder and St Clair 1978). The diodes allow a reduction in the complexity of external sensor addressing circuits by blocking parasitic conduction paths.

By using miniature components and hybrid circuit construction techniques, circuitry for scanning a sensor array can also be housed within the sensor enclosure. The circuit in Figure 3.11 uses a pair of cascaded Johnson counters to scan rows and columns of a 10 x 10 sensor array (Russell 1984).

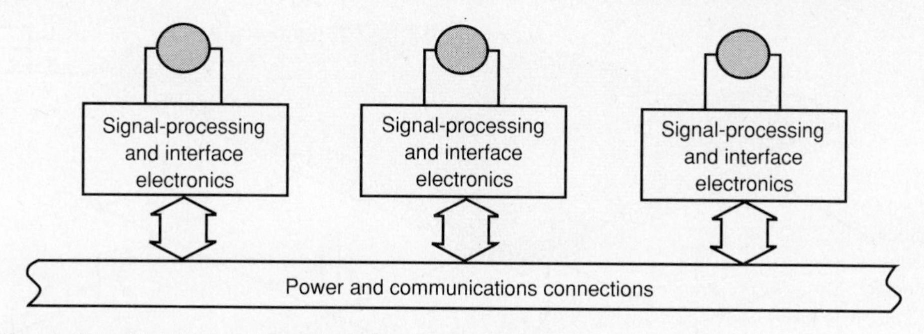


**Figure 3.9** Distributed computing with smart sensors

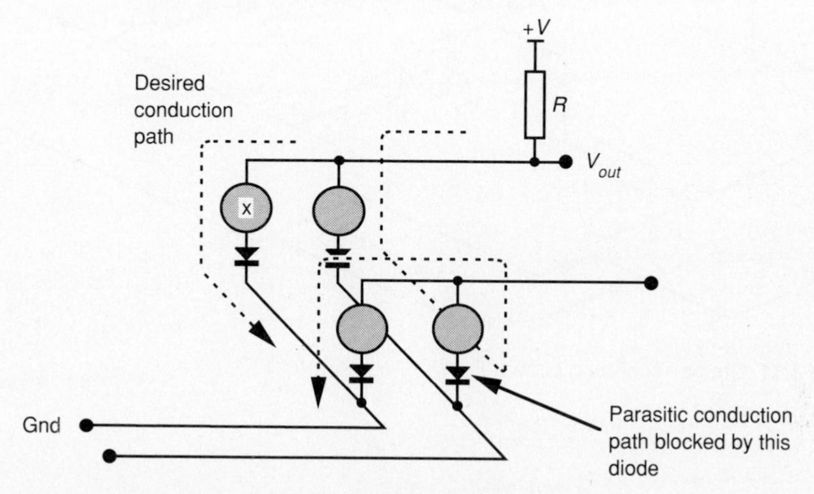


**Figure 3.10** Using diodes to block parasitic conduction paths

Sensor contact pads are connected to a directly coupled transistor AND gate so that each sensor pad can be 'pulled down' to ground in turn. The transducer effect used in this sensor is contact resistance which varies with normal force. The varying contact resistance forms part of a potential divider as shown in Figure 3.12. Resistance *R* is the fixed resistance between a point immediately above the contact area and the edge of the elastomer sheet where the common connection is made. This scanning circuit only requires five connections to the sensor.

Single chip microcontrollers are now available which contain parallel input/output lines, A/D (analog-to-digital) converters and UARTs (universal asynchronous receiver, transmitter) in addition to a complete microcomputer system. These devices have a great potential as the 'intelligence' for smart sensors.

## 3.6 Silicon sensors

Over the past twenty-five years a great deal of expertise has been accumulated in the fabrication of silicon integrated circuits. Both digital and analog circuit functions can be

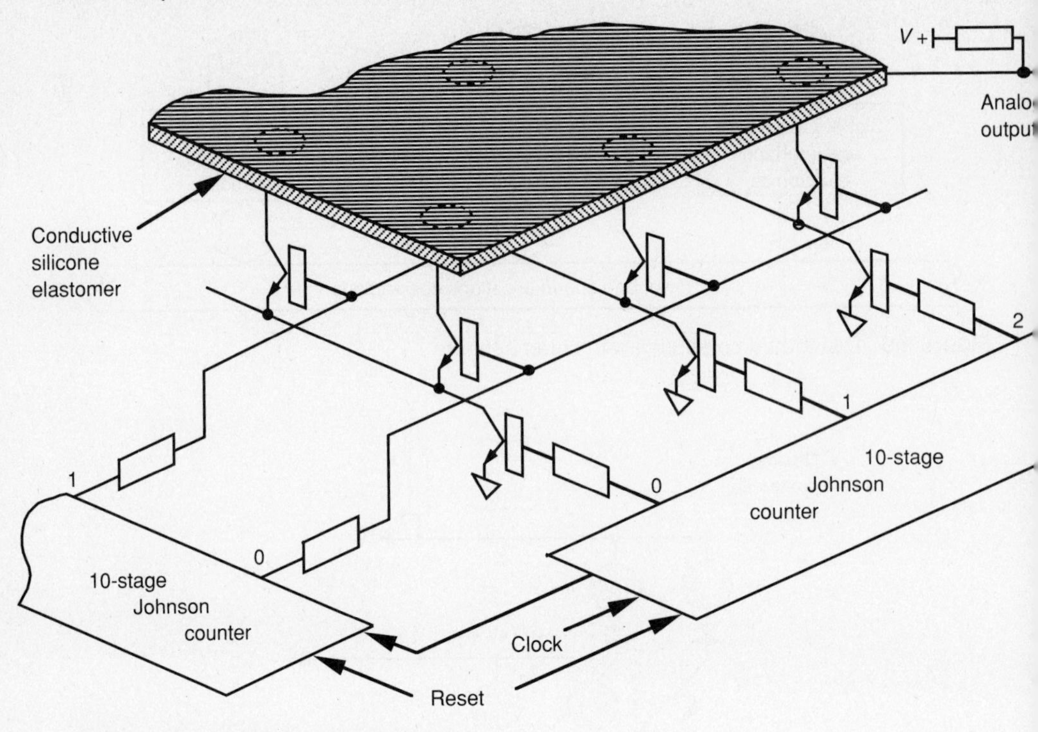


**Figure 3.11** Sensor-scanning circuit

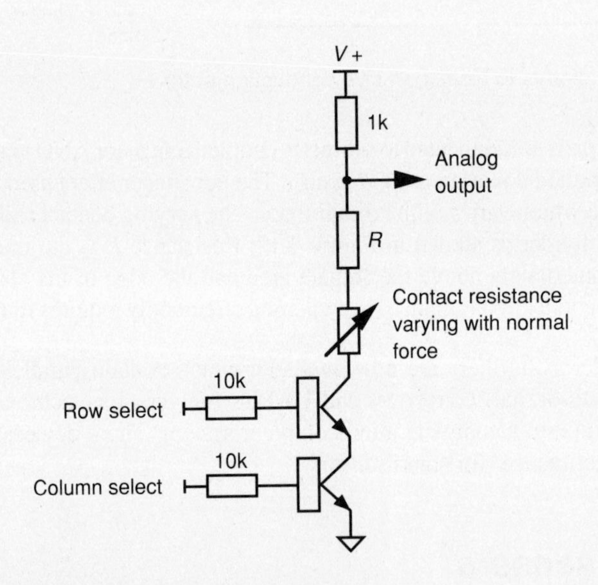


**Figure 3.12** Reading the contact resistance

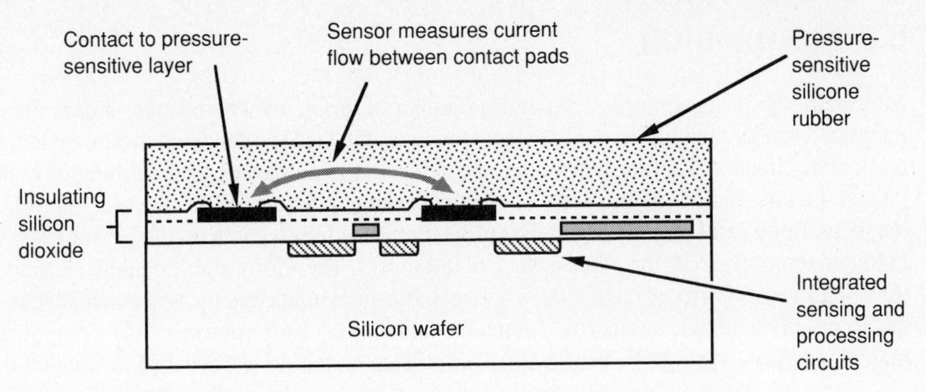


**Figure 3.13** Schematic cross-section view of a VLSI circuit tactile sensor

readily incorporated into the same piece of silicon. In addition, sensor devices can also be fabricated alongside circuit elements.

Silicon sensors employing the photoelectric effect (to measure light), Hall effect (to measure magnetic fields), and piezoresistance (to measure force or pressure) are available commercially, and many others are possible. Integration of circuit functions and sensor elements into a silicon sensor can yield many benefits (Middlehoek and Hoogerwerf 1985):

1. *Small size* — making them easier to use and reducing packaging costs.
2. *Fewer connections* — reduces the packaging cost and also increases reliability.
3. *Better signal-to-noise ratio* — weak signals can be amplified very close to where they originate. This eliminates noise pick-up caused by transferring low-level signals over long lengths of wiring.
4. *Internal compensation* — sensor non-linearity, offsets, and drift can be compensated by circuits integrated into the silicon sensor.
5. *Signal conditioning* — analog-to-digital conversion, data processing, and data reduction can be performed before the information is transferred from the sensor.

An early example of a tactile sensor design based on VLSI (very large scale integrated) circuit techniques (see Figure 3.13) was reported by Raibert and Tanner (1982). This device performs three functions:

1. *Transduction* — the surface of the VLSI chip contains pairs of electrodes which make contact with a layer of pressure-sensitive plastic (Dynacon B). The electrical resistance between the electrodes provides a measure of the applied force. A universal threshold is set for all the sensor elements, which then produce a binary output.
2. *Computation* — digital circuits can perform some simple data-processing operations on the data contained in each cell.
3. *Communication* — the output of all the sensor elements are transmitted, in serial form, to an output pad on the integrated circuit.

The VLSI-circuit-based tactile sensor reported by Raibert and Tanner also qualifies as an example of a smart sensor based on its computing and communications capabilities.

## 3.7 Conclusion

Addressing large sensor arrays presents a number of problems which smart and/or silicon integrated sensor techniques could solve. However, the tactile sensing environment can be particularly hostile. Tactile sensors may be subjected to damaging or disturbing environmental factors including impacts, abrasion, cutting, puncturing, heat, cold, humidity, magnetic fields, and electric fields. For protection the circuitry in a tactile sensor must be mounted remotely from the outer surface of the sensor. Therefore, attention must be paid to the reliability of electrical connections between sensor element and the associated electronics in an environment subject to stretching and flexing. Alternatively, the compliance matching approach could be employed. Compliance matching is described in Chapter 6.

## Bibliography

BALLANTYNE, D. W. G., and LOVETT, D. R., *A Dictionary of Named Effects and Laws in Chemistry, Physics and Mathematics*, Chapman and Hall, London and New York, 1980.

GIACHINO, J. M., 'Smart Sensors', *Sensors and Actuators*, Vol. 10, 1986, pp. 239-48.

LION, K. S., 'Transducers: Problems and Prospects', *IEEE Transactions on Industrial Electronics and Control Instrumentation*, Vol. IECI-16, No. 1, July 1969, pp. 2–5.

MIDDLEHOEK, S., and HOOGERWERF, A. C., 'Smart Sensors: When and Where?', *Sensors and Actuators*, Vol. 8, 1985, pp. 39–48.

MIDDLEHOEK, S., and HOOGERWERF, A. C., 'Classifying Solid-State Sensors: The "Sensor Effect Cube" ', *Sensors and Actuators*, Vol. 10, 1986, pp. 1–8.

RAIBERT, M. H., and TANNER, J. E., 'Design and Implementation of a VLSI Tactile Sensing Computer', *International Journal of Robotics Research*, Vol. 1, No. 3, Fall 1982, pp. 3–18.

RUSSELL, R. A., 'An Imaging Force Sensor for Robotics Applications', *Proceedings of the National Conference on Robotics*, Melbourne, 1984, pp. 123–7.

SNYDER, W. E., and ST CLAIR, J., 'Conductive Elastomers as Sensor for Industrial Parts Handling Equipment', *IEEE Transactions on Instrumentation and Measurement*, Vol. IM-27, No.1, March 1978, pp. 94–9.

## Questions

3.1 Briefly describe the three main components of an information-processing system.

3.2 Identify the six forms of energy that can carry information and for each of the five non-electrical forms of energy name a physical effect which converts it to an electrical signal.

3.3 If telepathy exists what can you say about the mechanism of information transfer?

3.4 Consider the following as information-processing systems:

(a) a radar-based proximity fuse in an anti-aircraft shell;
(b) a system controlling an electric pump to raise artesian well water into a dam and maintain a constant depth of water in the dam;
(c) a vacuum ignition advance system in a motor car engine; and
(d) a pocket calculator.

In each case identify the input and output transducers and the processing element. Identify the physical manifestation of information at each stage for each system.

3.5 In the context of sensors, explain the following terms:

(a) self generating;
(b) modulating;
(c) linearity; and
(d) resolution.

3.6 When addressing an array of sensor elements, under what circumstances would it be reasonable to use the following addressing schemes?

(a) two wires per sensor; and
(b) a potential divider chain.

3.7 A 4 x 4 tactile array sensor uses piezoresistive elements and row and column matrix addressing. Sensors are positioned at the cross-points of row and column wires. The wires are numbered 0 to 3 with row number followed by the column number. Sensor element (1,1) has a resistance of 500 Ω and elements (1,3), (3,1), and (3,3) each have a resistance of 1200 Ω. All other sensor elements have 5000 Ω resistance. What will be the resistance measured between the row (1) line and column (1) line if:

(a) the conditions are as stated above;
(b) element (1,1) has 5000 Ω resistance and the other elements are as stated in the question; or
(c) elements (1,3), (3,1) and (3,3) have 5000 Ω resistance and the other elements are as stated in the question.

*Note:* in each case all other row and column lines are open circuit while lines row (1) and column (1) are being tested.

3.8 Describe one technique for eliminating parasitic conduction paths in row and column addressing. What would be the effect if the interconnecting wiring had a significant resistance compared to the resistance of the sensor elements?

3.9 What is a 'smart' sensor?

# A kinesthetic sense for robots

Usually, tactile sensors will be mounted on a robot hand, arm, or some other mechanism to provide mobility. A kinesthetic sense measures two pieces of information which are important for a mobile tactile sensing system:

1. Localization (position and orientation) of the sensing system. When physical contact is established between a tactile sensor and an external object it is important to be able to calculate the exact position of the point of contact in world co-ordinates. This information localizes the object and allows a description of the object to be built up as a result of several contacts, independent of the varying position of the robot.
2. Sensor force and torque. Useful information can be determined by measuring the forces and torques resulting from contact between the robot and an external object. This information may be used to:

   (a) limit contact forces for safety reasons;
   (b) counter jamming and detect error conditions during assembly operations; and
   (c) measure material friction, compliance, pliability, and resilience.

   To provide this information we require methods of measuring the linear or rotational motion of the robot articulations and contact forces developed between the robot and the external object.

## 4.1 Measuring the position of robot joints

### 4.1.1 The potentiometer

Potentiometers are mechanically variable potential dividers. They consist of a resistive track and a movable wiper contact which samples the potential at a point along the track. If a voltage $V_{supply}$ is applied across the resistive track then the potential on the wiper $V_{output}$ is a function of the rotary or linear position of the wiper contact (see Figure 4.1).

For a rotary potentiometer, the angle $\theta_l$ that the shaft has turned through is given by:

$$\theta_l = \frac{R_l \theta_T}{R_T} = \frac{V_{output} \theta_T}{V_{supply}} \tag{4.1}$$

and for a linear potentiometer, the linear displacement $L_l$ of the wiper is:

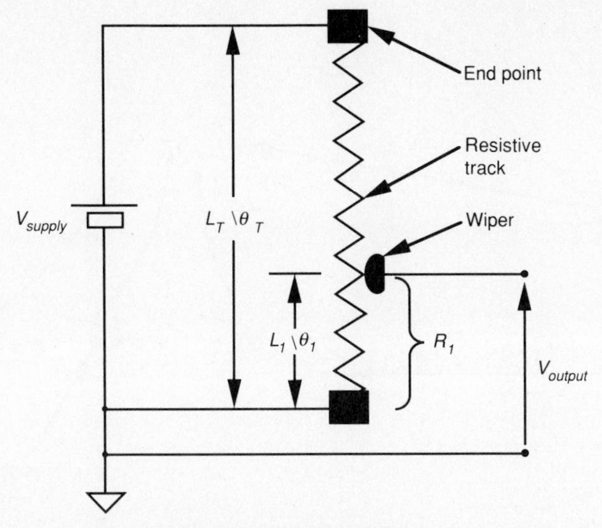


**Figure 4.1** Schematic diagram of a potentiometer

$$L_1 = \frac{R_1 L_T}{R_T} = \frac{V_{output} L_T}{V_{supply}} \tag{4.2}$$

where:

$R_T$ = total resistance of the resistive track between endpoints;
$R_1$ = resistance between the reference endpoint of the potentiometer track and the slider contact;
$\theta_T$ = total rotary travel of the shaft between endpoints; and
$L_T$ = total linear travel of the wiper between endpoints.

The resistive track may be made of a thin film of resistive material such as carbon, metal, conductive ceramic, or conductive plastic. An alternative is the wire-wound potentiometer where the resistive track consists of fine, high resistance wire wound onto an anodized aluminum former. As illustrated in Figure 4.2 the output of wire-wound potentiometers changes in small steps as the wiper jumps from one turn to the next whereas the resistive film form of potentiometer has a smooth variation in output. These step changes in the output of wire-wound potentiometers set a limit to their resolution.

Equations 4.1 and 4.2 are only correct if no current is drawn from the wiper contact. Therefore, the output of a potentiometer should be fed into a high-input impedance amplifier. For example, if the amplifier loading is to introduce an error of less than 0.1 per cent then the amplifier input impedance must be greater than 500 times the resistance of the potentiometer track. Potentiometer reliability will also be improved if minimal current is drawn through the wiper contact.

The output voltage of a potentiometer can be used directly to provide position information for an analog control system. Alternatively the voltage can be digitized by an analog-to-digital converter for processing by a computer.

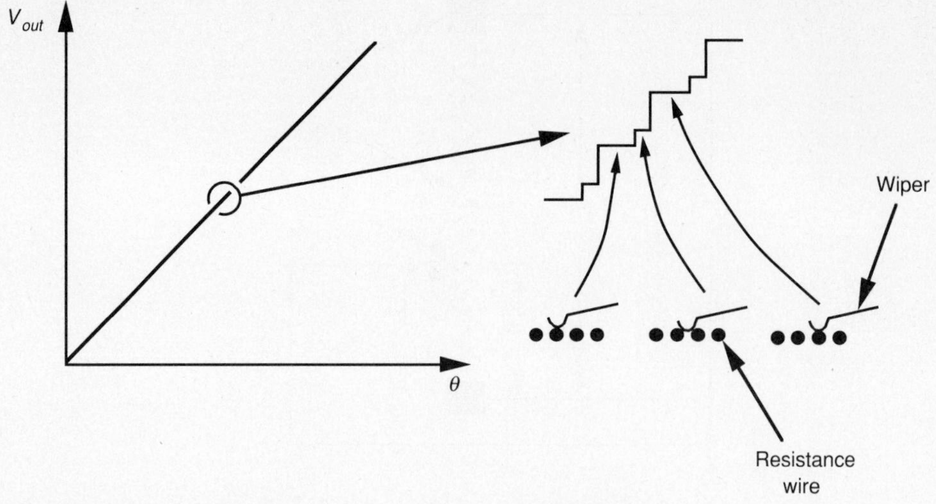


**Figure 4.2** Steps in the output of a wire-wound potentiometer

Potentiometers are available to measure linear movement and either single-turn (where the angle between endpoints is limited to about 300° because space is required for the end-point contacts) or multi-turn rotary motion. The advantage of potentiometers is that they provide absolute position information. They are also relatively cheap unless high accuracy is called for. Good potentiometers can have linearity errors below 0.1 per cent. In robotics potentiometers are used to measure joint angles in grippers and educational robots. However, potentiometers do not have the accuracy for very demanding applications and because they employ mechanical contact their characteristics change with time due to dirt, corrosion, and wear. Where a potentiometer is used to provide position feedback for a servo system, mechanical contact between wiper and resistive track introduces friction which adversely affects control of the servo system.

## 4.1.2 Optical encoders

Optical encoders enable a linear or rotary displacement to be converted directly into digital form without intermediate analog-to-digital conversion.

*4.1.2.1 Optical-slotted vane rotary switch*

The simplest form of optical encoder is the optical-slotted vane rotary switch shown in Figure 4.3.

A beam of light traveling between a light source (incandescent light bulb or light-emitting diode) and a light detector (silicon photodiode) is interrupted by an opaque slotted vane. As the shaft rotates light is alternately allowed through to the detector and cut off. By counting the number of dark-to-light and light-to-dark transitions seen by the photodetector, and knowing the number of slots in the vane, the resulting rotation of the shaft can be calculated. Figure 4.4 shows a simple circuit for providing a positive-going pulse for every dark-to-light and light-to-dark transition at the photodetector.

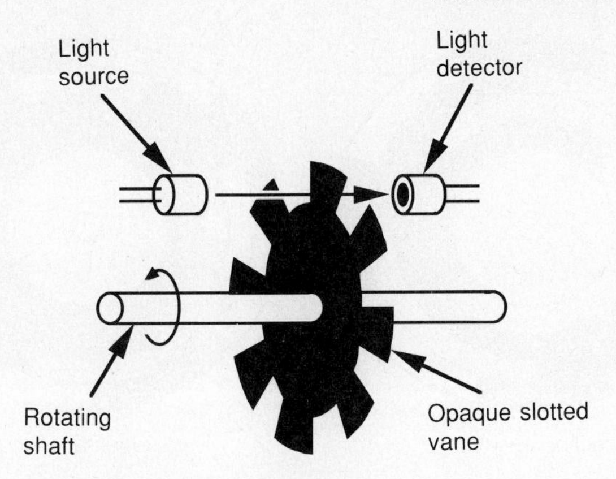


**Figure 4.3** The optical-slotted vane rotary switch

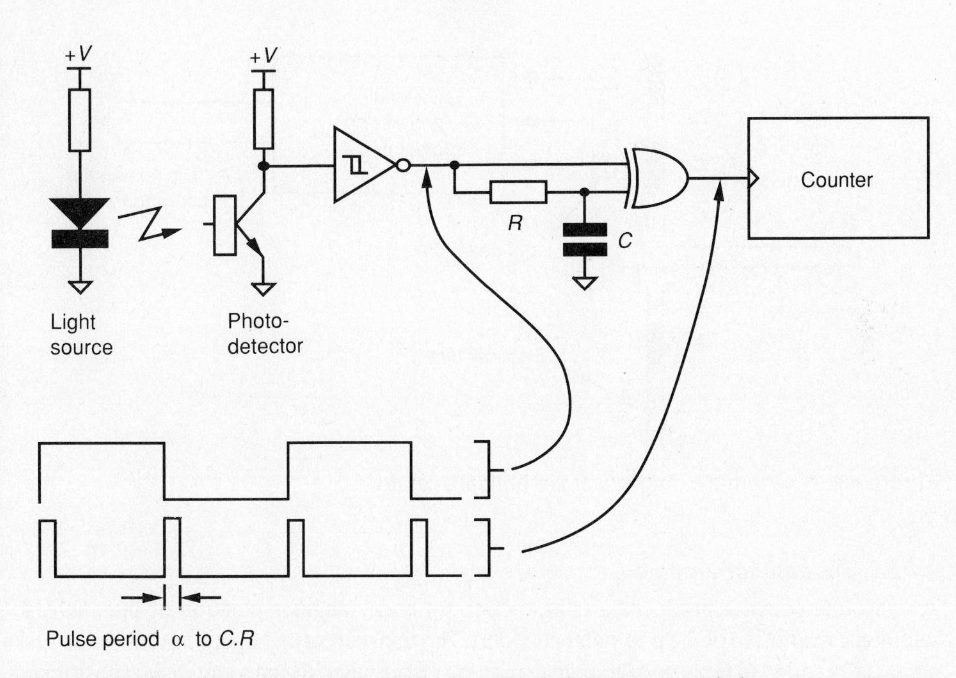


**Figure 4.4** Signal-conditioning circuit for an optical-slotted vane switch

This simple rotary switch does not give information about the direction of shaft rotation and can only be used where the direction is both constant and known. The incremental and absolute optical encoders overcome this restriction (see Figure 4.5).

The encoder input shaft is attached to a transparent disk of glass or plastic imprinted with areas of opaque material arranged in concentric tracks. Each track has a light source shining through the disk towards a corresponding light detector on the opposite side (see Figure 4.6).

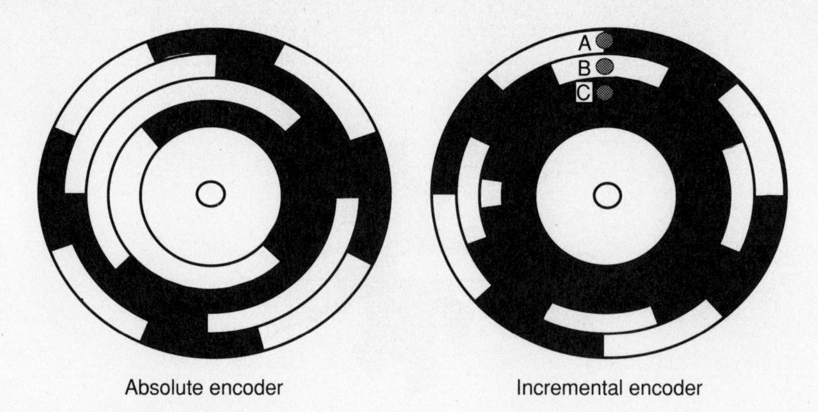


**Figure 4.5** Optical encoder disks

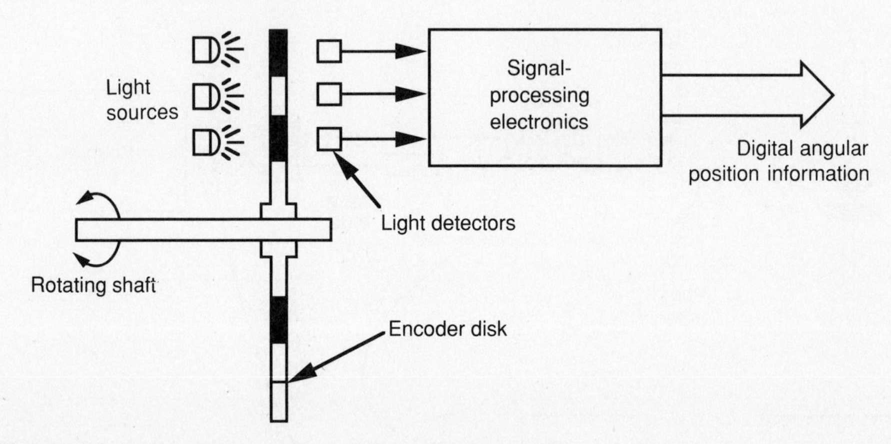


**Figure 4.6** A schematic diagram of the optical encoder

### *4.1.2.2 The absolute encoder*

An absolute encoder produces a unique code for every angular position of the shaft (to within the resolution of the particular encoder). The transparent and opaque areas on the disk are usually coded in Gray code to reduce reading errors when the encoder disk is positioned midway between two codes. If the disk is coded in binary, large errors can occur at these points on the disk. For example, when the encoder disk moves between tracks 0 and 15 then four bits change at the same time. Because of mechanical misalignment the bits will change at slightly different times. Therefore, depending on the exact order in which the bits change it is possible to read any combination of the four bits and large errors can result. Gray code only changes one output at a time and therefore large reading errors cannot occur. A comparison of binary and Gray code is shown in Table 4.1.

The simple combinational logic circuit shown in Figure 4.7 can be used to convert Gray code to binary.

The angular resolution of an absolute optical encoder depends upon the number of light

**Table 4.1** A comparison of binary and Gray code

| *Decimal* | *Natural binary* | *Gray code* |
|---|---|---|
| 0 | 0000 | 0000 |
| 1 | 0001 | 0001 |
| 2 | 0010 | 0011 |
| 3 | 0011 | 0010 |
| 4 | 0100 | 0110 |
| 5 | 0101 | 0111 |
| 6 | 0110 | 0101 |
| 7 | 0111 | 0100 |
| 8 | 1000 | 1100 |
| 9 | 1001 | 1101 |

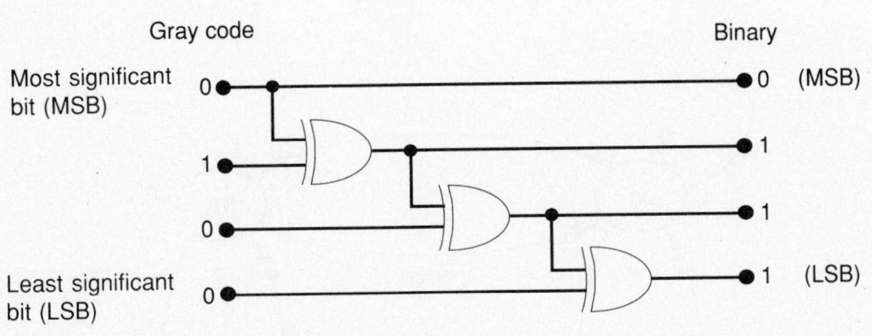


**Figure 4.7** Circuit to convert Gray code to binary

emitter/detector pairs and associated optical tracks on the encoder disk. An encoder with ten tracks produces $2^{10} = 1024$ different codes for one revolution of the encoder disk.

*4.1.2.3 The incremental shaft encoder*

The incremental encoder has only two or three light emitter/detector pairs and is much simpler and therefore cheaper than an absolute encoder of similar resolution. However, it requires more support circuitry and must be initialized whenever power is switched on. Referring to Figure 4.8, as the encoder disk rotates, the two sensors, A and B, produce square wave signals separated by a phase angle of 90°. If A leads B by 90° the disk is rotating anticlockwise and if B leads A by 90° the disk is rotating clockwise.

Sequential logic circuits (or a computer program if it can run fast enough to ensure that no input transitions are missed) can decode the outputs A and B and track movements of the encoder disk. The state diagram shown in Figure 4.9 decodes the output of the optical switches A and B and produces two outputs U and D which can control an up/down counter to track the movements of the incremental encoder. An asynchronous sequential circuit implementation of this state diagram is shown in Figure 4.10.

The incremental encoder needs to be initialized when power is switched on. An index track C produces a pulse once every revolution and this can be used to initialize the hardware.

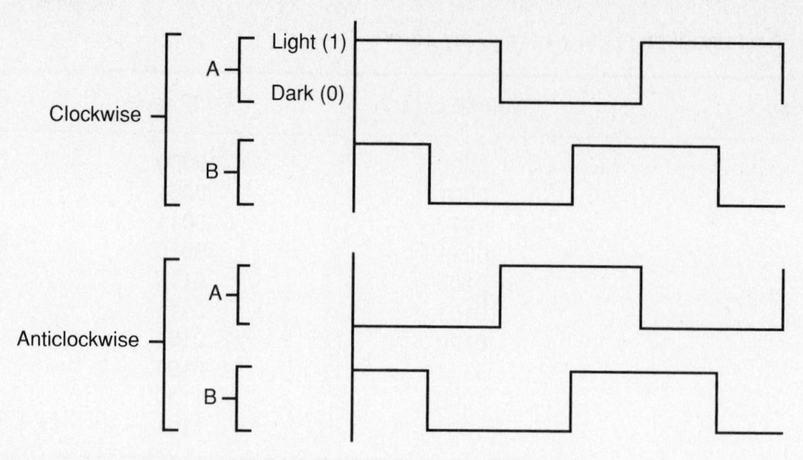


**Figure 4.8** Two-phase output from an incremental encoder

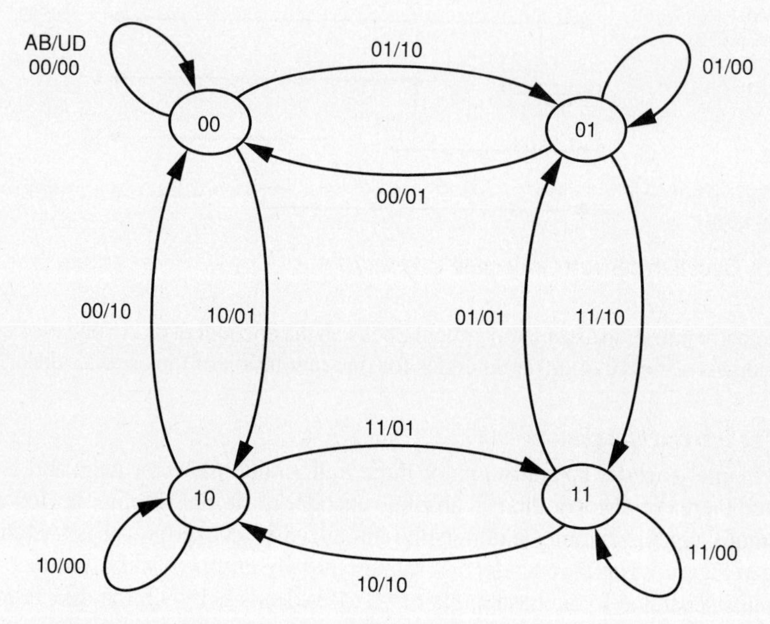


**Figure 4.9** State diagram of a sequential logic circuit to decode the output of an incremental encoder

### *4.1.2.4 Moiré fringes*

As we increase the resolution of an incremental encoder, the number of clear and opaque segments spaced around the encoder disk increases. To accommodate the extra segments the sensitive area of the photodiodes must be reduced (then the segments can be made smaller) or the disk diameter increased (in which case more segments of the same size can be accommodated).

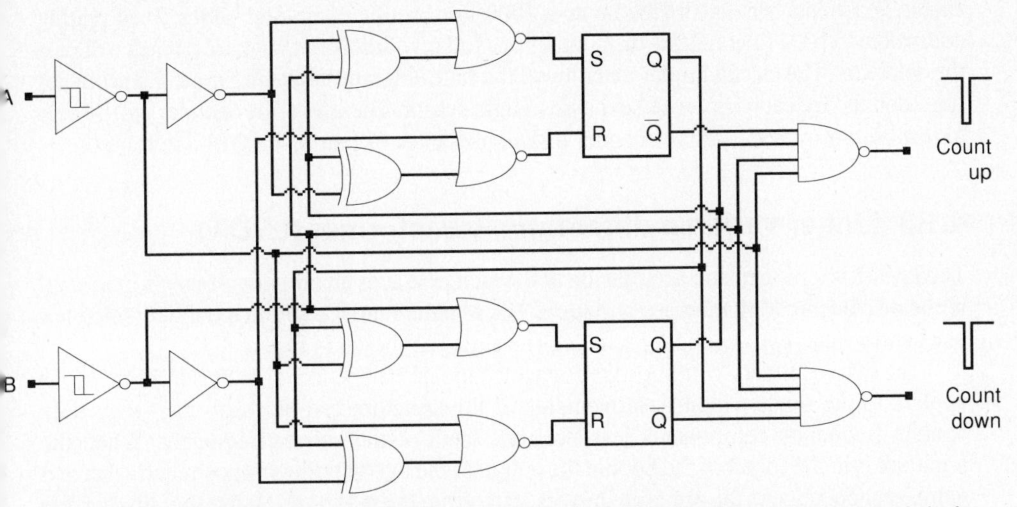


**Figure 4.10** Asynchronous sequential circuit to decode the output of an incremental shaft encoder.
(Courtesy of G. W. Trott and B. H. Smith, Wollongong University, Australia)

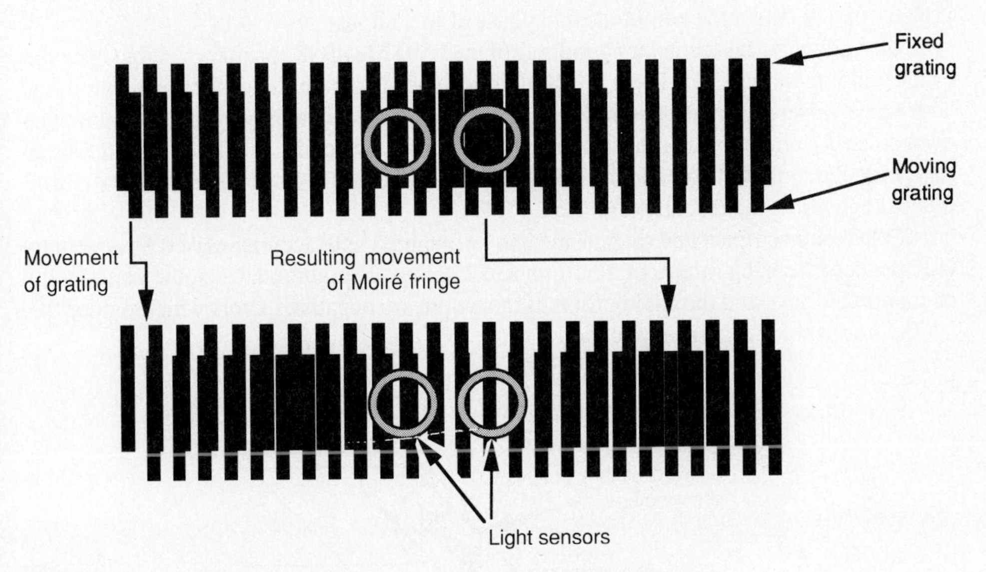


**Figure 4.11** Using Moiré fringes to measure linear or rotary motion

By using Moiré fringes a large increase in resolution can be obtained without reducing the sensitive area of the photo diodes or increasing the encoder diameter. The encoder disk is marked with finely spaced radial lines and between the disk and photodiodes is another transparent plate also marked with lines parallel to those on the disk but with a slightly different spacing. As shown in Figure 4.11 a small movement of the grating produces a large movement of the Moiré fringes.

*Example:* Consider a disk scribed with a 1000-line grating combined with a fixed grating equivalent to 1001 lines. As the disk rotates one full revolution, 1000 Moiré fringes will pass the detectors. The circuit implementation of the state diagram shown in Figure 4.9 produces four 'counts' for each fringe which passes a light sensor. Therefore, the encoder will be able to resolve a movement of the encoder disk equivalent to 1 part in 4000 of a revolution.

### 4.1.3 Linear variable differential transformer (LVDT)

The LVDT is a rugged and accurate device which produces an output voltage proportional to the displacement of a ferrous armature. The armature moves within a bobbin which has one primary and two secondary windings arranged as shown in Figure 4.12.

The two secondary windings are connected in series opposition. A carrier frequency is applied to the input winding. Movements of the armature can be resolved if they only contain frequency components less than one-tenth of the carrier frequency. When the armature is in the center of the bobbin the output of the two windings oppose each other and almost cancel out. As the armature moves away from the null position the flux linking one secondary increases while that linking the other decreases. An output voltage is produced, the phase gives the direction of armature movement and the amplitude gives the magnitude of movement (Figure 4.13). Support circuitry is required to produce a signal to excite the primary and to detect the amplitude and phase of the output.

The voltage in the secondary windings of the LVDT leads the primary voltage because of the winding inductance. There is a slight difference in the phase lead for each winding. This phase difference makes complete cancellation of the secondary voltages impossible even when the armature is at the centerline between the two coils. When the armature is at this point, the sum of the secondary voltages is called the null voltage and the magnitude of this voltage limits the resolution of the LVDT.

LVDTs are compact and rugged and can be obtained with accuracies of 0.5 per cent to 0.05 per cent. Sensing ranges of ±0.1 mm to ±250 mm are common. Cost, the complexity of support circuits, and the analog form of the output are negative factors which reduce the LVDT's appeal for robotic applications.

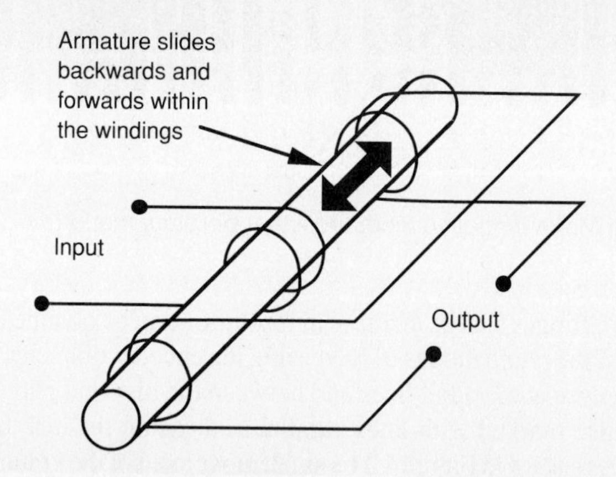


**Figure 4.12** Schematic diagram of the LVDT

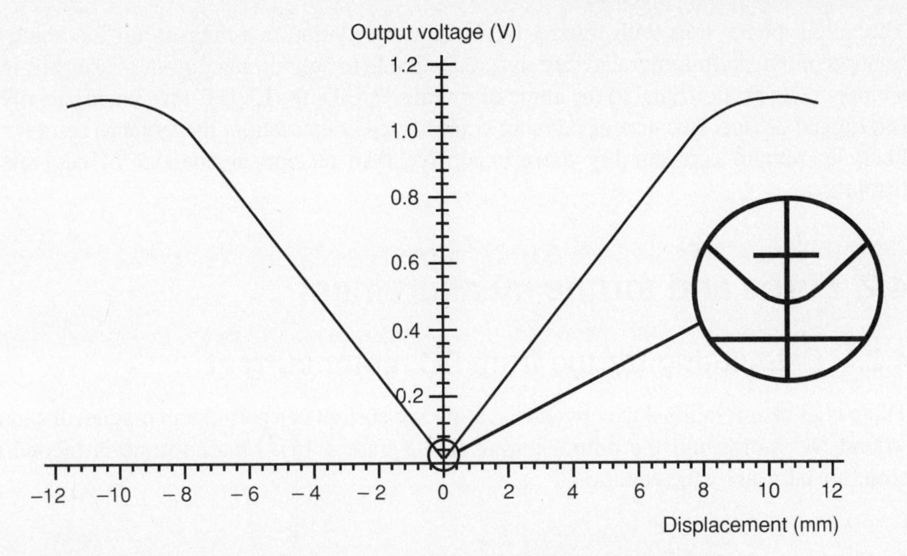


**Figure 4.13** LVDT output voltage as a function of armature displacement

## 4.1.4 Synchro resolver

A synchro resolver is essentially a variable coupling transformer. Two primary windings are positioned at 90° to each other on the stator and the rotor has a single winding (Figure 4.14).

The coupling between primary and secondary windings depends upon the position of the rotor. Quadrature signals:

$A = E \sin \omega t$ and $B = E \cos \omega t$

are fed into the two primary windings and the resulting secondary waveform is:

$$X = E \cos \omega t \sin \theta + E \sin \omega t \cos \theta \quad (4.3)$$

$$X = E \sin (\omega t + \theta) \quad (4.4)$$

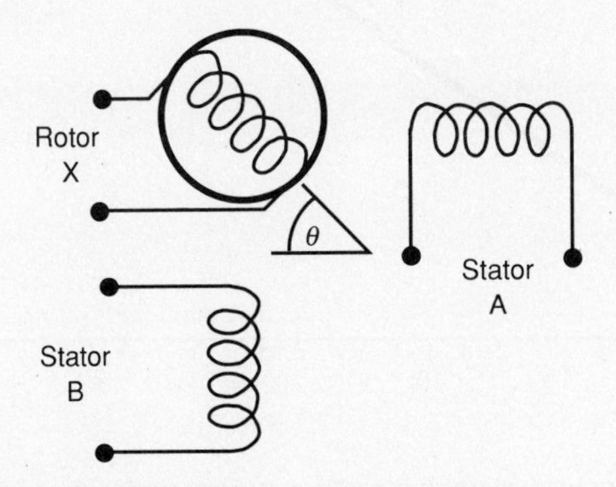


**Figure 4.14** Orientation of windings in a synchro resolver

The output phase shift with respect to the input excitation is a measure of the angle of rotation of the shaft. Integrated circuits are available to convert the output phase shift into a binary code proportional to the angle of rotation $\theta$. Like the LVDT, the synchro resolver is a rugged device. Because of its more complicated construction, the synchro resolver is likely to remain considerably more expensive than an optical encoder of equivalent resolution.

## 4.2 Force and torque measurement

### 4.2.1 Calculating torque from DC motor current

There is a linear relationship between the armature current in a permanent magnet or shunt-wound DC motor and the output torque (see Figure 4.15). Motor torque is related to armature current by the equation:

$$T_r = K \phi I_a \tag{4.5}$$

where:

$T_r$ = torque developed (newton meters);
$K$ = a constant for a particular motor;
$\phi$ = airgap flux/pole (webers); and
$I_a$ = armature current (amperes).

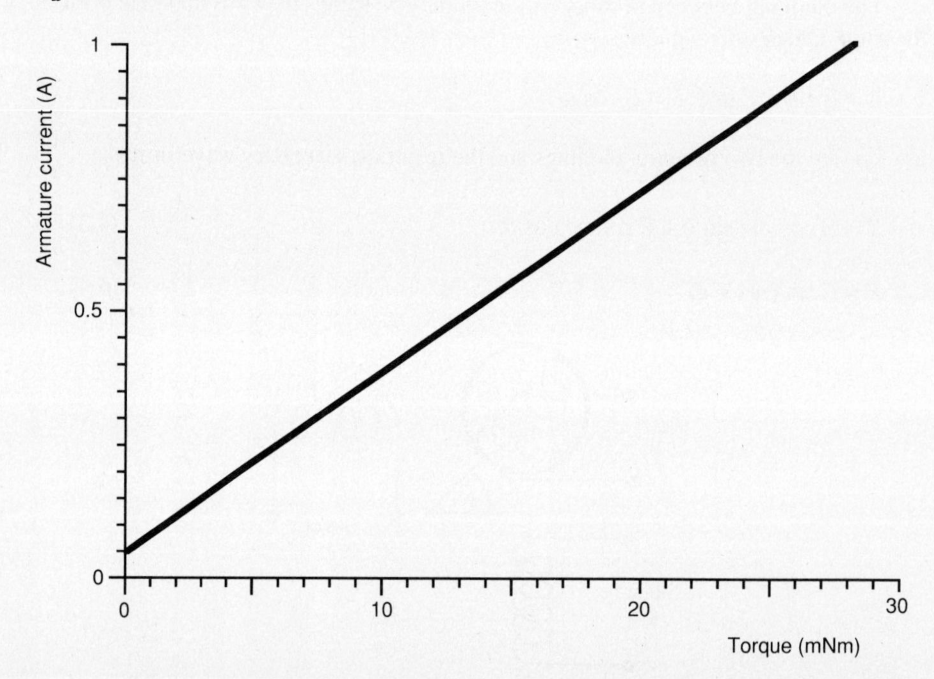


**Figure 4.15** The relationship between armature current and torque for a 3 watt permanent magnet DC motor

This law is obeyed reasonably well within the operating range of a motor. There is a discontinuity at very low current levels because stiction and friction prevent the motor moving until a threshold torque is achieved. Motors with iron in their rotors also suffer from cogging (caused by varying reluctance in the magnetic circuit), which is another factor preventing rotation at low torque levels.

Unfortunately, friction, stiction, backlash, and compliance in the drive train between motor and robot arm joint reduce the accuracy of the information obtained. Another problem is that a fair amount of calculation is required to account for the effects of gravity on all links of the robot and to allow for the varying configuration of the robot while calculating the forces and torques applied by the robot to an external object. The recent trend towards direct drive motors, eliminating gearbox and other drive chain components, should improve the accuracy of this technique.

## 4.3 Load cell structures

An alternative method of measuring forces and torques caused by the interaction of a robot with its environment is to measure the deflection of the robot structure. The links of most robot manipulators are made as rigid as possible and therefore there is little strain to be measured. The small amount of strain which does occur is distributed throughout the whole robot structure. The solution is to reduce the strength of the robot at one point with the aim of increasing and concentrating the resulting strain. Load cells are special deformable structures which localize and measure applied forces and torques. These are incorporated into the wrist or fingers of a robot manipulator. Because they represent a deliberately introduced weak point, load cells are susceptible to damage. The probability of damage can be reduced by:

1. end stops, which limit the deflection of the load cell elastic members to a safe value (see Figure 4.16); and
2. shear pins, made of a relatively low-strength material, which shear rather than transmit damaging stresses to the load cell (see Figure 4.17).

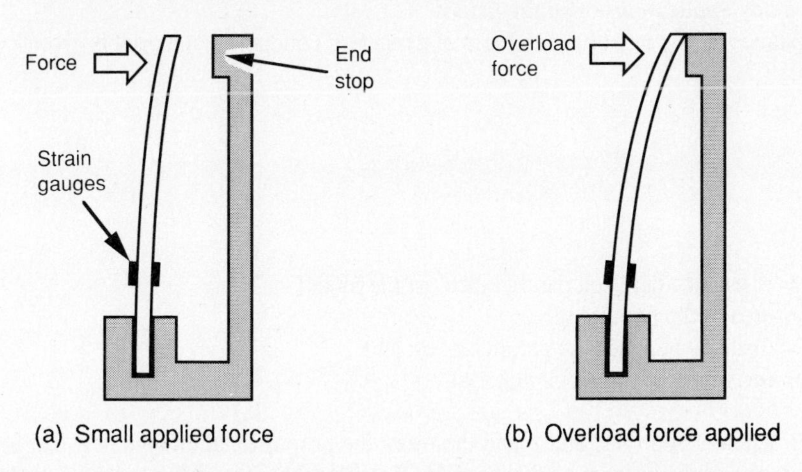


**Figure 4.16** Overload protection by using an end stop to limit deflection

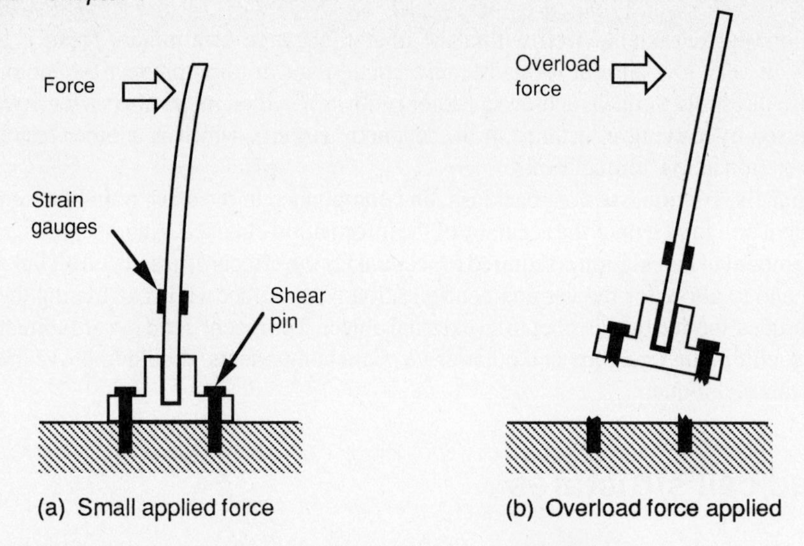


**Figure 4.17** Shear pins limit forces during overload

To obtain the maximum information three forces and three torques should be resolved, though some designs measure fewer parameters.

## 4.3.1 Strain gauges

Applied force cannot be measured directly but only by using mechanical elasticity to convert force into deflection which can be measured. Strain is the measure of deflection. If a piece of material has an original length $l$ and is then subjected to a force which changes its length by a small amount $\Delta l$ the resulting axial strain is defined as $\Delta l$ divided by $l$.

In 1856 Lord Kelvin described his observation that the resistance of some metal wires varied in proportion to extension as they were stretched. This piezoresistive effect is used today in the many applications of strain gauges.

The resistance across two opposite faces of a block of conducting material is given by the equation:

$$R = \frac{\rho l}{A} \tag{4.6}$$

where:

$R$ = resistance between parallel faces of the block;
$A$ = area of the parallel faces;
$l$ = distance between the parallel faces; and
$\rho$ = resistivity of the block material.

Resistance $R$ depends upon $A$, $l$ and $\rho$ and therefore the proportional change in $R$ can be expressed in terms of the proportional changes in the other three variables:

$$\frac{\Delta R}{R} = \frac{\Delta \rho}{\rho} + \frac{\Delta l}{l} - \frac{\Delta A}{A} \tag{4.7}$$

As a force is applied to the block its length will change and, as a result, so will its cross-sectional area. Poisson's ratio $\mu$ relates change in length to change in cross-sectional area:

$$\mu = \frac{transverse\ strain}{axial\ strain} \tag{4.8}$$

Therefore the relationship between a proportional change in length and the associated change in area is:

$$\frac{\Delta A}{A} = -2\mu \frac{\Delta l}{l} \tag{4.9}$$

Substituting Equation 4.9 into Equation 4.7 we get:

$$\frac{\Delta R}{R} = \frac{\Delta \rho}{\rho} + (1 + 2\mu)\frac{\Delta l}{l} \tag{4.10}$$

From Equation 4.10 we can see that change in resistivity and change in shape both contribute to the change in resistance of a piece of conductive material when it is stretched or compressed.

The first strain gauges were made from fine wires glued to an insulating backing. Many modern gauges are formed from a thin metal foil punched or etched into the required shape. Figure 4.18 shows a diagram of a metal foil strain gauge. The shape of the gauge is chosen to minimize transverse sensitivity. The sensitivity of strain gauges is expressed in terms of gauge factor $K$:

$$K = \frac{\Delta R / R}{\Delta l / l} \tag{4.11}$$

A list of typical gauge factors for metal strain gauges is given in Table 4.2.

**Table 4.2** Gauge factors of common strain gauge materials

| *Material* | *Composition (%)* | *Gauge factor (K)* |
|---|---|---|
| Chromel | Ni64, Fe25, Cr11 | +2.5 |
| Manganin | Cu84, Mn12, Ni4 | +0.5 |
| Nichrome | Ni80, Cr20 | +2.2 |
| Nickel | | −12 |
| Platinum | | +4.8 |

Semiconductor strain gauges (see Figure 4.19) are made from silicon or germanium doped with small quantities of donor (n) or acceptor (p) impurities. They exhibit a very large

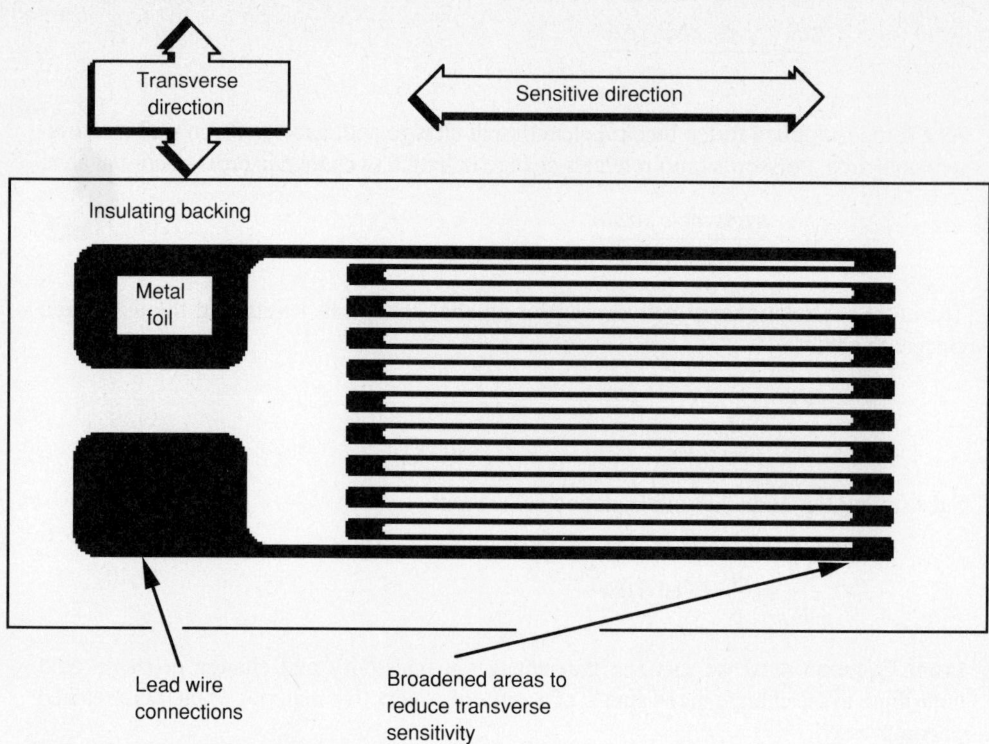


**Figure 4.18** A metal foil strain gauge

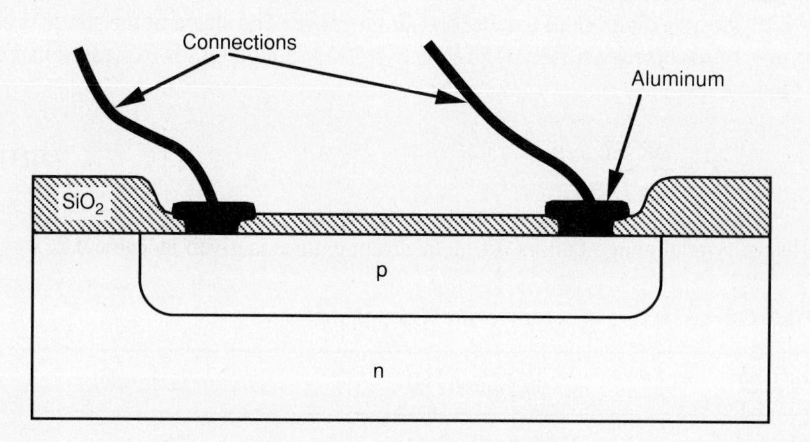


**Figure 4.19** A semiconductor strain gauge

variation of resistivity with strain and have a gauge factor of the order of 100 times that of metal gauges. However, semiconductor gauges are non-linear, temperature sensitive, brittle, and approximately ten times as expensive as foil gauges. Figure 4.19 shows a cross-section view of a p-type strain gauge. P-type impurities are diffused into a lightly n-doped wafer of silicon. The diffusion can follow a serpentine path to raise the resistivity of the gauge. Aluminum pads are deposited to make contact to the gauge and a layer of silicon dioxide forms a protective layer over the device.

Strain gauges formed in semiconductor materials offer the exciting prospects for including signal processing and communications circuits within the same piece of material.

In use, strain gauges are glued to some structure which is undergoing strain. The strain is transferred through the glue and backing material to the gauge element where it causes a change in resistance in the gauge element. The Wheatstone bridge illustrated in Figure 4.20 is a versatile circuit for measuring the change in resistance. If all arms of the bridge have resistance $R$ and then the element in branch A changes by $\Delta R$ then the resulting change in output voltage is:

$$\Delta V_{output} = \frac{V_{supply}}{2R} \cdot \frac{\Delta R}{(2 + \Delta R / R)} \tag{4.12}$$

Providing the change in the gauge resistance is small compared to the total gauge resistance then:

$$\Delta V_{output} = \frac{V_{supply} \Delta R}{4R} \tag{4.13}$$

Consider the cantilever beam in Figure 4.20. If gauge A only is used with B, C, and D fixed resistors of the same nominal value as A, then the output will be sensitive to changes in temperature which affect the gauge resistance. Using gauges in positions A and D provides temperature compensation. If gauges are used in positions A, B, C, and D then the bridge sensitivity is doubled and temperature variations in the gauge elements are compensated.

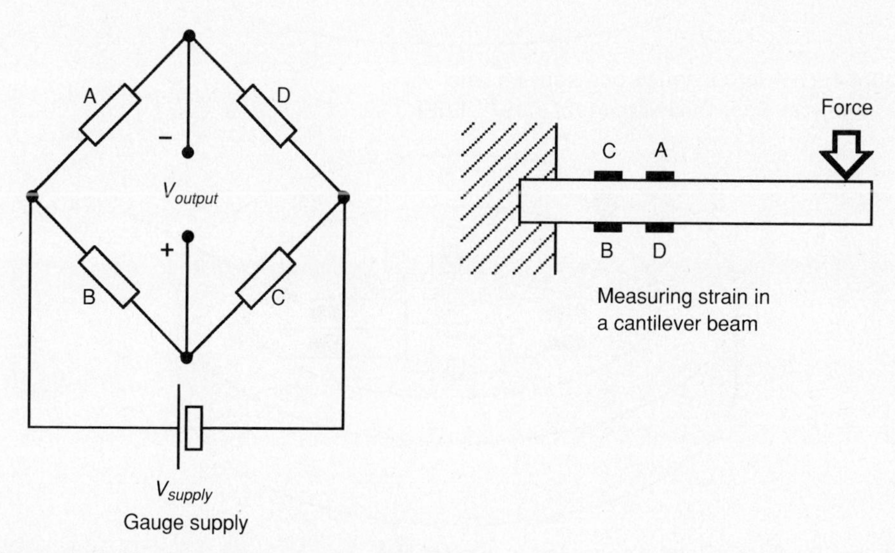


**Figure 4.20** The Wheatstone bridge connection

### 4.3.2 Load cell designs

Figure 4.21 shows a force- and torque-sensing wrist developed at Stanford Research Institute (SRI) for use on the Stanford arm (Rosen and Nitzan 1977). The intricate shape was machined out of a section of aluminum tube. Incorporated within the force and torque sensor were eight thin beams of aluminum orientated to detect forces and torques applied in three orthogonal directions.

A more compact force/torque sensor was made for the Jet Propulsion Laboratory (JPL), California Institute of Technology and reported by Bejczy (1980) (see Figure 4.22). This sensor was machined from a single block of aluminum and had the form of a spoked wheel.

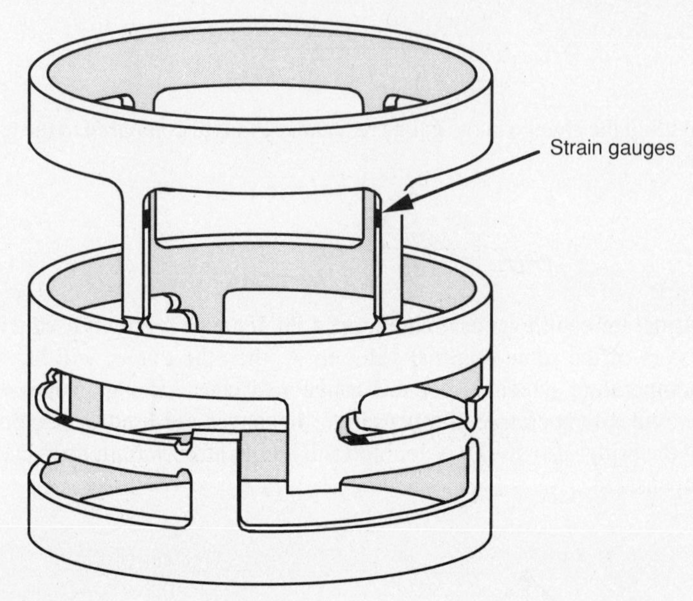


**Figure 4.21** A force- and torque-sensing wrist
(Redrawn from Rosen and Nitzan 1977, ©1977 IEEE)

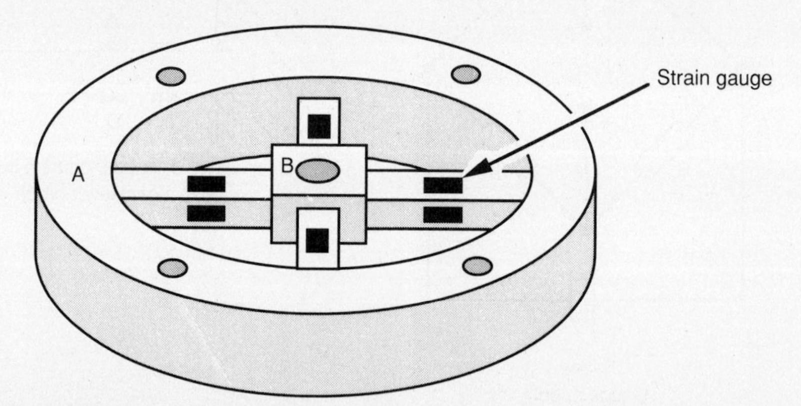


**Figure 4.22** A force/torque sensor developed at JPL
(From Bejczy 1980, ©1980 AIAA, reprinted with permission)

**Figure 4.23** A force-sensitive wrist measuring forces and torques acting on a robot gripper (Photograph courtesy of Lord Corporation, Industrial Automation Division, Cary, North Carolina)

The outer rim of the wheel A is connected to the robot arm and the inner hub B to the gripper (see Figure 4.23). There are four semiconductor strain gauges attached to each spoke of the wheel and they are wired in pairs as half-bridge circuits.

A miniaturized version of the spoked wheel load cell has been mounted within the finger of a robot hand to provide a direct measurement of forces and torques encountered by one finger. An alternative finger load cell has been described by Bicchi and Dario (1988). The form of this load cell is a thin walled tube with six strain gauges mounted on the outside (Figure 4.24). A seventh gauge attached to the mounting flange provides temperature compensation.

The JPL and SRI load cells produce eight output signals. The output voltages $V$ ($v_0$ to $v_8$) are related to forces and torques $F$ ($f_1$ to $f_6$) applied to the load cell by the following equation:

$$V = H \, . \, F \tag{4.14}$$

In terms of the individual components:

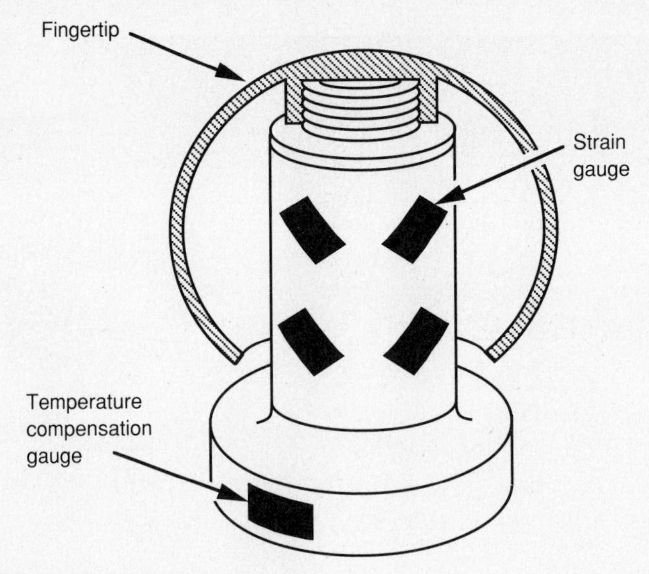


**Figure 4.24** A finger-mounted load cell

$$\begin{bmatrix} v_1 \\ v_2 \\ \cdot \\ \cdot \\ v_n \end{bmatrix} = \begin{bmatrix} h_{11} h_{12} \cdot\cdot h_{16} \\ h_{21} h_{22} \cdot\cdot h_{26} \\ \cdot \\ \cdot \\ h_{n1} h_{n2} \cdot\cdot h_{n6} \end{bmatrix} \cdot \begin{bmatrix} f_1 \\ f_2 \\ \cdot \\ \cdot \\ f_6 \end{bmatrix} \qquad (4.15)$$

The quantities $h$ are found either by calculating the effects of forces and torques on the load cell structure or by applying pure forces and torques and using the resulting output voltages to calculate the values of $h$. By inverting $H$ the values of applied forces can be calculated from the load cell output voltages.

Where the H matrix is non-square a pseudo-inverse $H^*$ can be used:

$$H^* = (H^t H)^{-1} H^t \qquad (4.16)$$

In the case of Dario's sensor the H matrix is square and therefore can be inverted directly.

## 4.4 Tendon tension sensing

With tendon-actuated multi-finger grippers the tension in each tendon can be monitored. The tendon is routed over an idler wheel supported by a cantilever beam. Tension in the tendon deflects the beam and this is detected by strain gauges (see Figure 4.25). Measuring

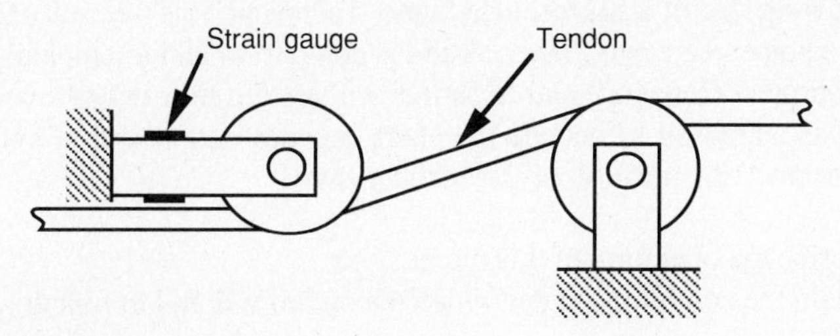


**Figure 4.25** Cable tension sensor
(Redrawn from Salisbury and Craig 1982, ©1982 IEEE)

tension in the tendon bypasses friction in the motor, gearbox and tendon sheath and provides the information required to calculate forces applied by the robot hand.

Many of the devices and techniques required to provide robot kinesthetic sensing have already been developed for other applications. The same is not true for the development of an artificial sensory skin. Applications for a skin-like sensor are limited to areas such as robotics and prosthetics. This presently small market and the very great technical problems involved have slowed down the development of skin-like sensors. The next two chapters describe some of the progress which has been made towards an artificial sensory skin for robots.

## Bibliography

BEJCZY, A. K., 'Smart Sensors for Smart Hands', *Progress in Astronautics and Aeronautics*', Vol. 67, 1980, pp. 275–304.

BICCHI, A., and DARIO, P., 'Intrinsic Tactile Sensing for Artificial Hands', *Robotics Research: The Fourth International Symposium*, R. Bolles and B. Roth, eds, MIT Press, Cambridge, Massachusetts, 1988, pp. 83–90.

OLIVER, F. J., *Practical Instrumentation Transducers*, Hayden, New York, 1971.

PERRY, C. C., and LISSNER, H. R., *The Strain Gauge Primer*, McGraw-Hill, New York, 1962.

ROSEN, C. A., and NITZAN, D., 'Use of Sensors in Programmable Automation', *Computer*, Vol. 10, No. 12, December 1977, pp. 12–23.

SALISBURY, J. K., and CRAIG, J. J., 'Articulated Hands: Force Control and Kinematic Issues', *International Journal of Robotics Research*, Vol. 1, No. 1, Spring 1982, pp. 4–17.

## Questions

4.1 A 10 kΩ rotary potentiometer with a travel of $0^{\circ}$ to $300^{\circ}$ between endpoints feeds into the 25 kΩ input impedance of an amplifier. If the potentiometer shaft is positioned at $100^{\circ}$, what angle does the resulting output voltage correspond to? (*Note:* the answer will be independent of the potentiometer supply voltage.)

4.2 The following circuit was presented in the 'Technical Tips' section of an electronics magazine some years ago. Inputs A and B come from an incremental shaft encoder and the outputs 'Count up' and 'Count down' are fed into an up/down counter. The circuit was supposed to decode the shaft encoder output and allow the up/down counter to track the position of the encoder shaft.

(a) Describe the operation of this circuit.
(b) Explain the conditions under which the circuit will fail to function correctly.

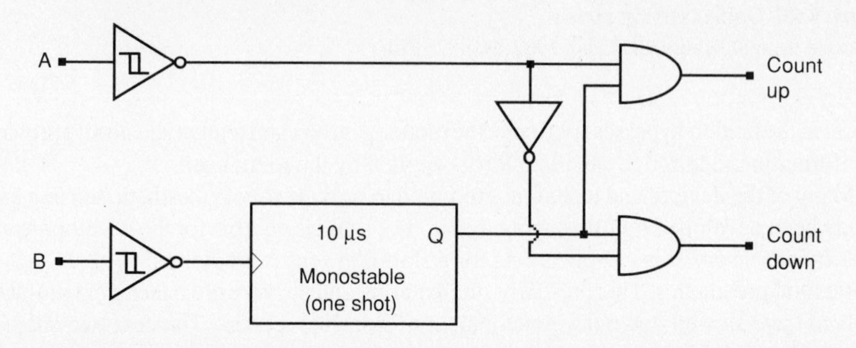


4.3 Compare the advantages and disadvantages of absolute shaft encoders and incremental shaft encoders.

4.4 Describe the operation of a linear variable differential transformer (LVDT).

4.5 Explain the advantages and disadvantages of measuring the forces and torques applied by a robot arm to an external object by measuring the current in its DC servo motors.

4.6 Describe any methods you can think of for measuring the position of the end effector of a robot without having to calculate it from the lengths of the robot links and the joint angles. The methods must be accurate (about ±0.1 mm) and not restrict movement of the robot.

4.7 The gripper of a robot arm is positioned 1 m radially out from its base and moves tangentially 0.1 mm. An incremental optical encoder is directly connected to the base rotate axis (i.e., rotating the base $10^\circ$ also rotates the encoder $10^\circ$). How many pulses per revolution must the encoder produce in order that the 0.1 mm tangential movement produces one recognizable change in output for the encoder (on average)?

4.8 A metal rod has an axial length of 100 mm. A strain gauge with a gauge factor of 2.2 and nominal resistance 250 Ω is attached to the rod to measure axial strain. An axial tensile load is applied to the rod. The single strain gauge is connected into a Wheatstone bridge circuit with a $V_{supply}$ of 5 V. $V_{output}$ of the bridge circuit changes by 0.01 V. What is the new length of the rod?

4.9 Describe how one, two, and four strain gauges can be used to measure the strain in a cantilever beam. If a Wheatstone bridge connection of the gauges is used, explain the advantages of using two and four gauges compared to a single gauge.

4.10 For the Wheatstone bridge circuit of Figure 4.18 show that if gauges A and D both change by a small amount $\Delta R$ (perhaps due to a change in temperature) there will be no variation in $V_{output}$.

4.11 Explain why a load cell must be protected against mechanical overload and suggest suitable techniques.

# 5 Touch sensor arrays

This chapter surveys some of the many different technologies which have been used to make touch sensor arrays. Interest in the area of robot tactile sensing goes back to the early 1960s with the work of Heinrich Ernst (1962). The idea of building a mechanical hand controlled by a computer was originally proposed by Shannon and Minsky during a seminar at the Massachusetts Institute of Technology in 1958. Based on this idea Ernst built 'MH-1', the first computer-controlled mechanical hand incorporating tactile sensing. The two-fingered gripper carried by MH-1 incorporated both on-off touch switches and small arrays of variable resistance sensors giving an output proportional to gripping force. Today, after more than twenty-five years of development, tactile sensor systems are approaching a level of capability which makes them useful for robotic applications. However, presently available sensors are still far from ideal. What are the requirements for a robot touch sensor? In 1980 Leon Harmon conducted a survey of fifty-five people from industry and research organizations in an attempt to answer this question (Harmon 1982). One of the results of this survey was a list of desirable performance specifications for a robot tactile sensor:

- Spatial resolution 1–2 mm (this is the approximate minimum separation at which the human fingertip can distinguish two points applied to the skin as separate stimuli).
- Array size between 5 x 10 and 10 x 20 points per fingertip (the human fingertip contains approximately 10 x 15 touch-sensitive points).
- Threshold sensitivity about 0.5–10 g for one force-sensing element.
- Dynamic range 1000:1.
- Stable, monotonic, and repeatable sensor response with no hysteresis.
- Sampling rate between 100 Hz and 1 kHz.
- Broadly — the sensor to be skin-like, rugged, and inexpensive.

These specifications have been used as a guide by many sensor designers. However, the touch sensory requirements for a particular application may differ in some respects from those listed above.

## 5.1 Force, pressure, and shear

If a constant force were applied to the sensitive region of a force sensor, then the output of the sensor would be constant. Sensor output would not be affected by changing the area of contact or the distribution of force over that area. By contrast, a pressure sensor output, for constant applied force, would depend inversely upon the area of contact. In this case the distribution of force over the contact area would affect the sensor response. Most of the devices considered in this chapter have a response which falls between that of a force sensor and pressure sensor. For this reason I have chosen to group them under the generic title of touch sensors.

The tendency for an object to slip from the grasp can be monitored by determining shear forces acting tangentially to the skin surface. This information will be important for monitoring and controlling slip. Some of the sensor designs reviewed in this chapter can be modified to make them sensitive to shear forces.

## 5.2 Switches as touch sensors

A switch is one of the simplest touch-actuated devices for detecting the presence of an object. Inside a 'normally open' switch two electrical contacts are initially separated. A certain threshold force applied to one of the electrodes bends it into contact with the other electrode and the resulting flow of current indicates that the threshold force has been exceeded. An array of commercially available miniature switches could form an inexpensive tactile sensor. However, even miniature switches are quite large, and therefore the array would have a low spatial resolution. This limitation, coupled with the binary (only signalling 'on' or 'off') nature of the output signal, severely limits the quantity of information that such an array can provide.

Further development of the idea of using switch arrays has produced a number of interesting and potentially useful sensors.

### 5.2.1 Pneumatic touch sensor

Some computer keyboards and good quality calculators have keys with 'feel' which give an audible and tactile click when depressed. In many cases the click is produced by a shallow spherical dome made of thin sheet metal.

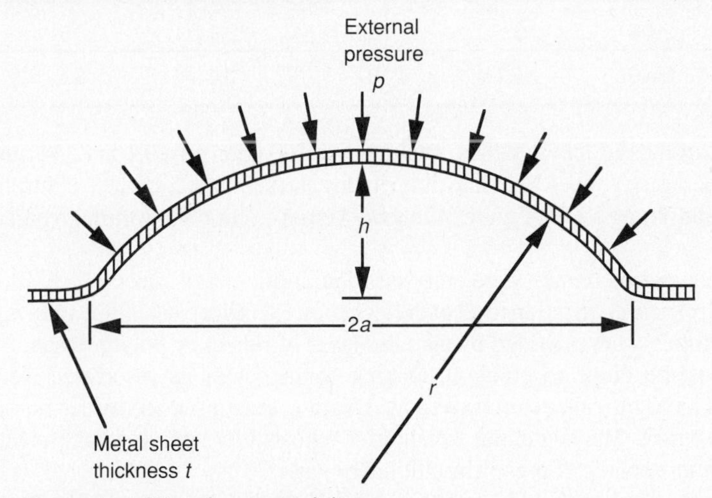


**Figure 5.1** Geometry of the sheet metal dome
(From Garrison and Wang 1973, redrawn with permission of IBM Corporation, Armonk, New York)

When a critical value of force is applied the dome collapses (with a click), and later returns to its original shape as the force is removed. The behavior of thin metallic domes under an applied load is predicted by the theory of elastic buckling (Libove 1962). Referring to the cross-section diagram of a shallow spherical dome shown in Figure 5.1, the geometrical parameter $\lambda$ is given by:

$$\lambda^2 = \frac{a^2}{t \cdot r}\sqrt{12(1-\upsilon^2)} \tag{5.1}$$

and the load parameter $R$ by :

$$R = \frac{p}{E}\left(\frac{a}{t}\right)^4 (1-\upsilon^2) \tag{5.2}$$

where:

$p$ = the externally applied pressure;
$E$ = Young's modulus for the metal sheet material; and
$\upsilon$ = Poisson's ratio for the metal sheet material.

For the snap switching action to occur, $\lambda$ must have a value between 2.08 and 6. When $\lambda$ is less than 2.08, the dome deforms continuously (no click) and for $\lambda$ greater than 6 it buckles in a more complicated manner.

Pressure $p$ required to cause snap switching of a dome can be found from Equation 5.2. Equations 5.1 and 5.2, and Table 5.1 can then be used to choose a suitable material and geometry for the dome.

**Table 5.1** Values of geometrical parameter $\lambda$ and associated critical load parameter $R_{cr}$ corresponding to the start of snap switching of the dome.

| $\lambda$ : | 2.08 | 3 | 3.5 | 4 | 5 | 6 |
|---|---|---|---|---|---|---|
| $R_{cr}$ : | 2.46 | 4.1 | 8 | 14 | 33 | 77 |

Arrays of these dome switches, totaling over 100 units on a 0.1 in (2.54 mm) row and column spacing, have been built into the gripping surfaces of a computer-controlled gripper (Garrison and Wang 1973). Figure 5.2 shows a cross-section view through part of a sensor array.

The snap-action domes were embossed into a thin metal sheet which also formed a common electrical connection to all of the switches. Electrical insulation and a high-friction gripping surface were provided by an outer layer of rubber or polyurethane. An external force causes the dome to collapse, making contact with its associated electrode and completing an electrical circuit. As an added feature, sensor sensitivity can be varied under computer control. The switching threshold is adjusted by applying back pressure to the switches from a source of pressurized liquid or gas.

Arrays of switches could be interconnected for row and column addressing. However, additional components, such as a diode in series with each switch, would be required to

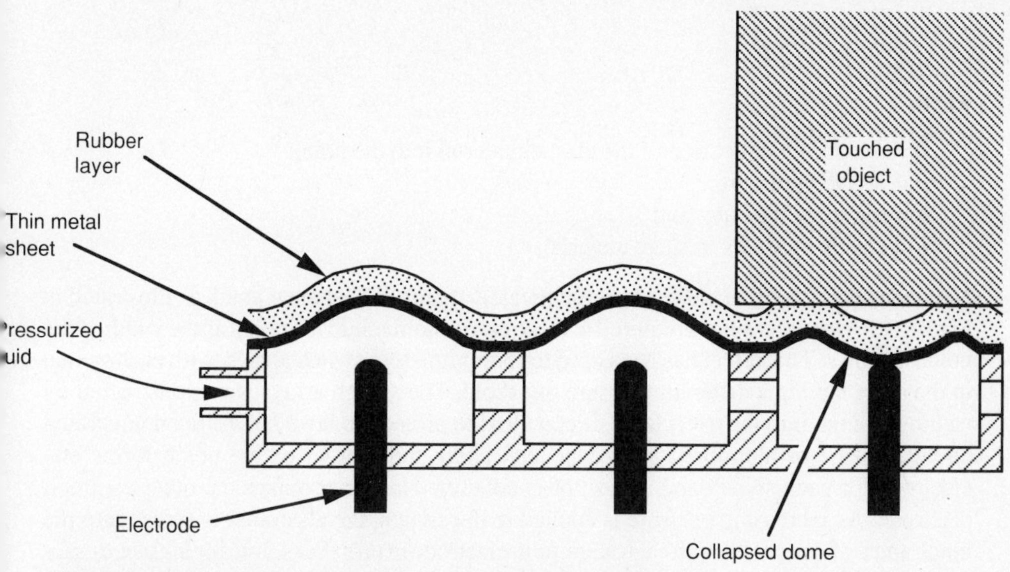


**Figure 5.2** Cross-section view of the pneumatic touch sensor
(From Garrison and Wang 1973, redrawn with permission of IBM Corporation, Armonk, New York)

eliminate parasitic conduction paths. The pneumatic touch sensor was reported in 1973 and, apparently, has not been developed further. Possible difficulties with this sensor include:

- physical constraints which make it difficult to decrease the spacing between switch elements which would improve spatial resolution;
- if the force applied to the dome has a shear component, this may prevent the dome from buckling with the required snap action;
- thin metal sheet domes may be susceptible to damage;
- the switch elements exhibit substantial hysteresis; and
- the binary output of the switch gives much less information compared to a sensor with continuously varying output.

## 5.2.2 Digital tactile sensor array

A simple switch indicates whether or not the applied pressure/force exceeds a set threshold. If a closely grouped array of switches could be made, each having a different threshold, then the magnitude of an applied pressure or force could be estimated. Conventional switches would be unsuitable because they are too large and their switching threshold cannot be adjusted.

A tactile sensor array based on this idea has been constructed using VLSI technology (Raibert 1984). If a sheet of elastic material is pressed against a round hole the material bulges into the hole. For small deflections, the maximum depth that the material protrudes into the hole will be related to the applied pressure, size of the hole and elastic modulus of the sheet material:

$$\delta \propto \frac{pa}{E} \tag{5.3}$$

where:

$\delta$ = maximum deflection of the elastic material into the hole;
$p$ = applied pressure;
$a$ = radius of the hole; and
$E$ = elastic modulus of sheet material.

Now consider a sheet of elastic material pressed against a V-shaped notch as illustrated in Figure 5.3. More pressure is required to force the elastomer into the notch as the width of the notch narrows. This effect has been used to form miniature switch arrays, with each switch in the array having a different pressure threshold. The switch arrays were constructed by etching a notch into the overglass (silicon dioxide protective layer) of a silicon integrated circuit. A row of aluminum pads deposited along the bottom of the notch forms one electrode for each switch and a sheet of conductive elastomer makes the other common electrode. As increasing pressure is applied to the switch, the elastomer is forced into the notch and makes contact with each aluminum electrode in turn. Sensor-scanning electronics are integrated into the silicon wafer to sample the state of each switch and transfer this information from the sensor as a serial sequence of bits.

A prototype version of the sensor has been reported which contains forty-eight notches, each located in a 0.3 mm by 0.6 mm area of the chip. The fifteen switch contacts would allow the sensor to discriminate sixteen pressure levels. Unfortunately, this prototype did not work reliably because of surface irregularities in the conductive elastomer sheet. The designer hoped to remedy this problem by increasing the thickness of the overglass and hence deepening the notch.

This sensor design has the advantage that it does not require an analog-to-digital converter to prepare data for a digital computer. Linear, logarithmic, or exponential response to pressure can be obtained by varying the shape of the notch. A disadvantage of this and some other silicon-based tactile sensors is that the fragile silicon chip is close to the point of contact with external objects. In this position, the sensor will be prone to damage. To a large extent, the surface of the silicon chip defines the shape of the sensor. It will be difficult to mold the sensor round the compound curves of a fingertip and the sensor sensitive area is limited to the maximum size of integrated circuit which can be fabricated.

## 5.3 Piezoresistive

### 5.3.1 Conductive elastomers

Conductive elastomers are insulating natural or silicone-based rubbers made conductive by adding particles of conducting or semiconducting materials (e.g., silver or carbon). These materials are a very popular starting point for the construction of tactile sensors. They are relatively inexpensive, easy to use, and their flexibility provides a good gripping surface with a significant amount of compliance. Several grades of conducting rubber are available commercially for use in electromagnetic shielding and as flexible electrical contacts in

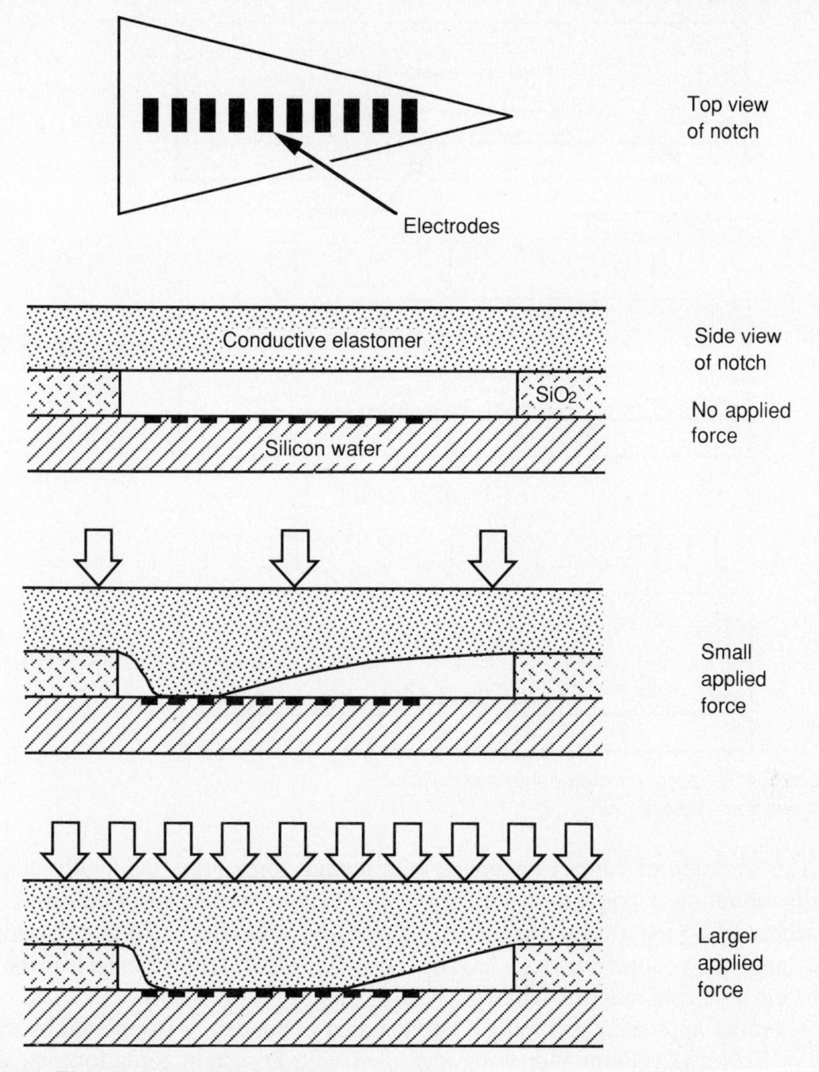

**Figure 5.3** The all-digital VLSI tactile array sensor
(Redrawn from Raibert 1984, ©1984 IEEE)

small electronic assemblies such as digital watches. Most of these forms of conductive rubber show little change in bulk resistance as they are compressed. However, area of contact and hence inverse contact resistance can be made to vary with applied force.

The sensor described by Hillis (1981) uses this principle to sense normal force. A separator (the woven mesh of a nylon stocking) gives no contact, hence infinite resistance, for zero normal force. At a certain threshold force the conductive elastomer squashes through the separator and makes contact with an electrode. Additional force increases the area of contact and thus reduces the contact resistance. This process is illustrated in Figure 5.4.

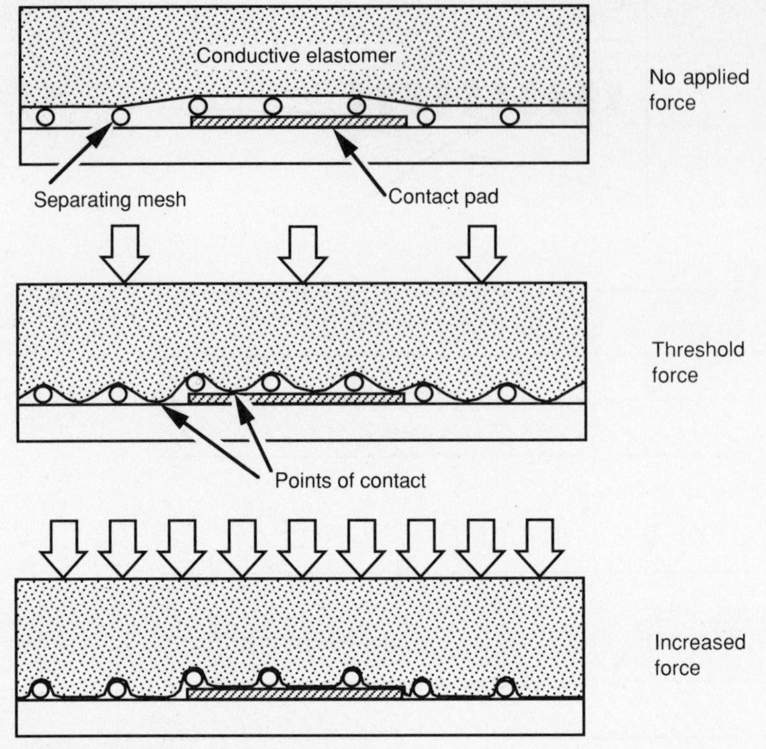


**Figure 5.4** Piezoresistance using a separator
(Redrawn from Russell 1985)

The variation of contact resistance with normal force gives the highly non-linear, rapidly saturating response shown in Figure 5.5. In Hillis' sensor, anisotropically conducting rubber ACS (zebra strip) was used to provide row electrodes and printed circuit tracks the column electrodes. This sensor has an array of 256 tactile sensor elements in the area of 1 $cm^2$ and a sensing range of 1–100 g.

A similar approach, also using a varying area of contact, was reported by Purbrick (1981). Row and column addressing was used with D-section cords forming the row connections and either printed circuit traces or another group of D-section cords for the column connections. Increasing normal force distorts the D-section, increasing contact area and decreasing contact resistance between rows and columns (see Figure 5.6).

Normal force plotted against contact resistance for this sensor follows the same general trend shown in Figure 5.5. However, the absence of a spacer means that there is still a finite contact resistance for zero applied force.

Several kinds of conductive rubber have been formulated to give a large change in bulk resistance when compressed. For example, Dynacon Industries manufacture a number of pressure-sensitive silicone rubber materials consisting of a silicone rubber matrix containing a dispersion of metallic particles coated with a semi-conducting and deformable layer of organo-metallic gel. Snyder and St Clair (1978) built a 4 x 4 sensor array using them. They found that although the Dynacon materials were easy to use and performed adequately as touch sensors, problems of material fatigue caused the sensors to wear out quickly.

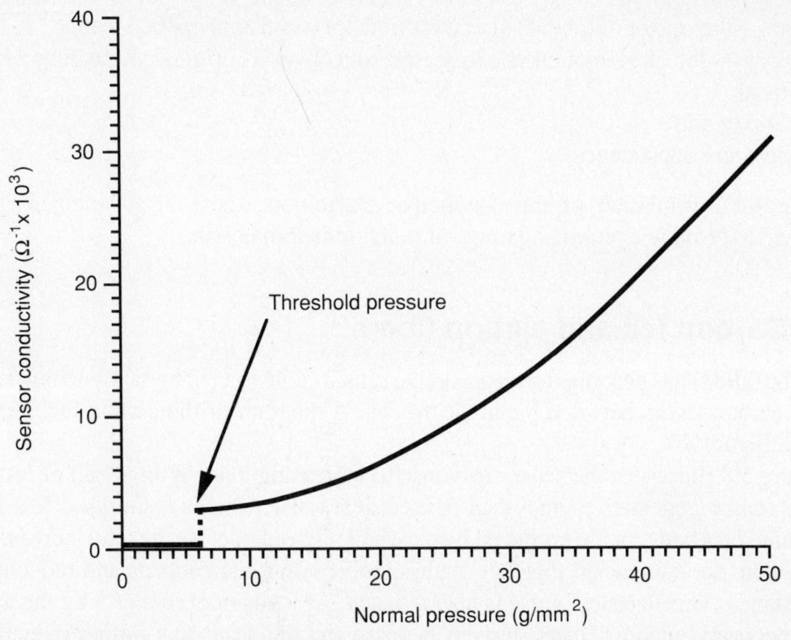


**Figure 5.5** Normal force plotted against contact resistance for a sensor using conductive rubber with a separator mesh

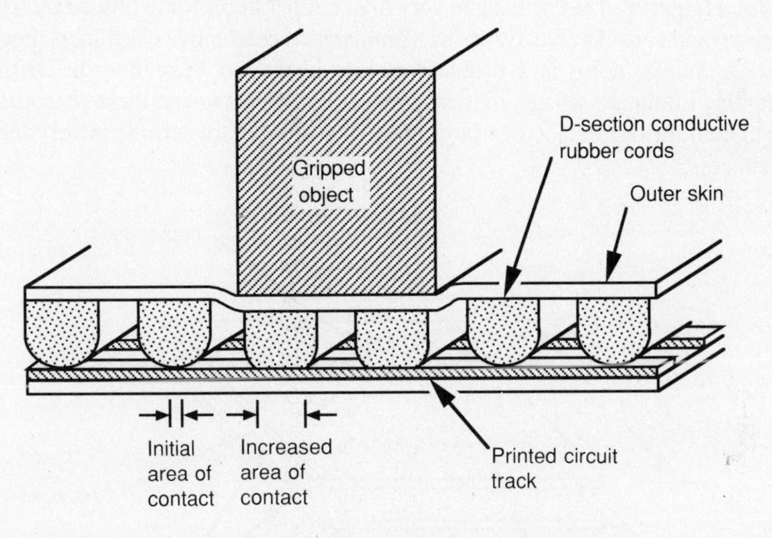


**Figure 5.6** Piezoresistance using D-section conductive rubber cords

Silicone-based conductive elastomers are chemically inert and withstand a wide range of temperatures. They can be molded into the compound curves required for a fingertip shape and provide a good gripping surface. Most sensor designs based on conductive elastomer produce a large electrical output signal with a frequency response ranging from DC to tens or hundreds of herz. Their disadvantages include:

- *creep* — the sensor output changes over time for a constant applied load;
- *memory* — the elastomer takes a long time to recover its original shape after a load is removed;
- *hysteresis*; and
- *temperature* dependence.

With careful formulation of the conductive elastomers, these disadvantages can be minimized to provide a promising range of touch sensor materials.

### 5.3.2 Carbon felt and carbon fibers

Larcombe (1981) has described piezoresistive sensors constructed by sandwiching carbon felt and carbon fibers between metal electrodes. A diagram of the carbon felt sensor is shown in Figure 5.7.

Figure 5.8 illustrates the sensor response to increasing load. With a load of less than 10 g the sensor generates a great deal of electrical noise. This is a similar effect to the background hiss and crackle produced by a carbon microphone. As the load increases, the carbon fibers are compacted together, making more electrical contacts and reducing the felt resistance. The electrical noise is also reduced. At loads in excess of 5 kg the area of contact between touching fibers starts to increase and this leads to a further reduction in resistance.

Carbon fiber and carbon felt sensors are rugged and can be shaped to conform to the curves of a fingertip. They withstand very high temperatures (only limited by oxidation of the carbon) and considerable overloads. Compared to conductive elastomers they have low hysteresis. Sensor noise is a problem at low loads and there may be difficulties in constructing miniature sensors for dense sensor arrays. However, these sensors seem well suited to monitoring contact over large areas of a robot or for sensing in very inhospitable environments.

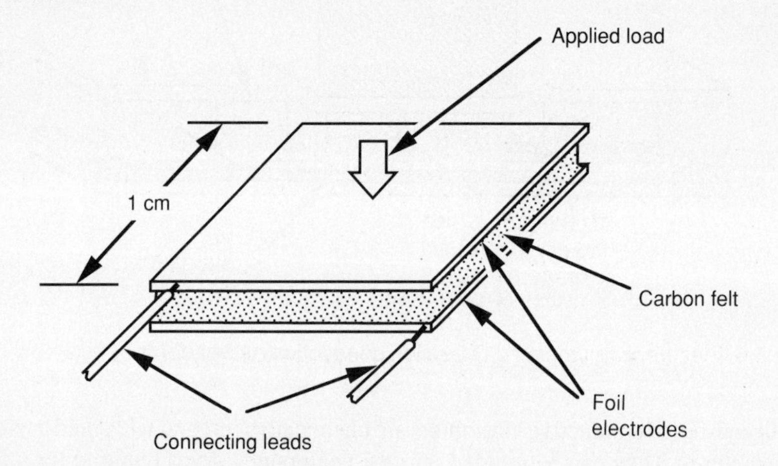


**Figure 5.7** Carbon felt tactile sensor
(Redrawn from Larcombe 1981, courtesy of IFS Publications and the author)

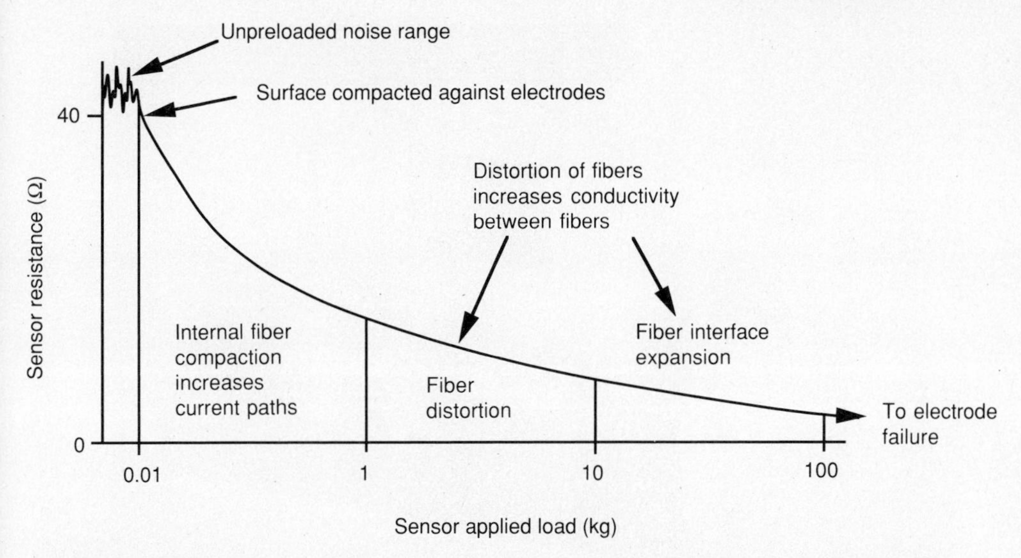


**Figure 5.8** Graph of load against resistance for a 1 cm by 1 cm carbon felt touch sensor (Redrawn from Larcombe 1981, courtesy of IFS Publications and the author)

## 5.3.3 Semiconductor strain gauge

Both metal and semiconductor piezoresistive elements have been used to construct tactile sensor arrays. Conventional metal foil strain gauges are rather large for this application, especially when they are bonded to a deformable element to convert the applied force into a strain for the gauges to measure. Semiconductor techniques allow very small strain gauge elements to be fabricated. The same semiconductor material used to fabricate the strain gauges can also form the deformable element, thus combining these two functions in the same piece of material. In addition, signal-conditioning electronics can be built into the same piece of silicon (Petersen et al. 1985).

Figure 5.9 shows a silicon touch sensor manufactured by IC Sensors of Milpitas, California. The sensor is made of two parts, a silicon chip bonded to a glass substrate. Both the silicon chip and glass substrate form mechanical components of the sensor and are micro-machined to produce the required shape. Using integrated circuit construction techniques, strain gauging, and addressing, circuits are formed on one face of the silicon wafer. Areas of the other face are machined away to concentrate forces in the region of the strain gauges. The glass substrate is etched to provide recessed areas into which the silicon chip can be deflected and parts of the glass substrate are metalized to provide electrical connections for the sensor. These two components are bonded together and a protective layer of silicone rubber is added to make the complete sensor.

The response of this kind of silicon touch sensor is very linear, especially when compared to most of the alternative touch-sensing techniques. They also exhibit low hysteresis and creep. Multiplexing, linearizing, and temperature compensation circuits can be incorporated into the silicon chip. Although overload protection is built into the sensor, it is still vulnerable to massive overload. The intrinsically planar nature of silicon integrated

**Figure 5.9** A silicon touch sensor
(Photograph courtesy of IC Sensors of Milpitas, California)

circuits presents a problem when fingertip-shaped sensors are required. The cost of the materials in a silicon integrated circuit is very low, in the range of cents per square millimeter, while the design and prototype stages of production cost tens or hundreds of thousands of dollars. Therefore, the price of silicon integrated circuits is very dependent on how many are sold. At the moment, the market for silicon integrated sensors is relatively small and hence they are quite expensive, in the range of tens to hundreds of dollars each.

## 5.4 Piezoelectric polymers

The piezoelectric effect was discovered by Jacques and Pierre Curie in 1880. They found that crystals of quartz produce an electrical charge when pressure is applied to the crystal.

The piezoelectric effect only occurs in crystals which do not have a center of symmetry. Such crystals have a dipole moment which changes with applied stress. If a piezoelectric material is left open circuit, as stress is applied then a voltage develops across the crystal followed by a reverse voltage as the stress is removed. Alternatively, providing a short circuit current path causes charge to flow in one direction as the stress is applied and in the reverse direction as the stress is removed. For quartz, a stress of 1 N/m produces a polarization of about 2 $pC/m^2$. Some recently discovered polymers based on polyvinylidene fluoride (PVF2) show the largest piezoelectric effect of any known material. Their flexibility, small size, sensitivity, and large electrical output offer many advantages for sensor applications in general and touch sensors in particular.

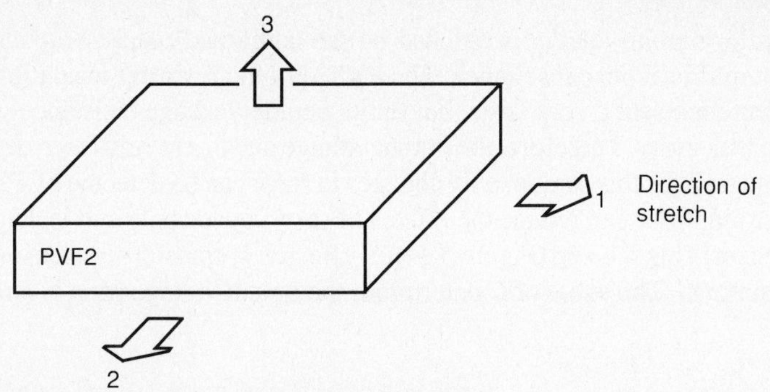


**Figure 5.10** Reference axes in PVF2 sheet material

To manufacture piezoelectric film, PVF2 is extruded into a sheet of the desired thickness and then controlled stretching orientates the polymer chains in the plane of the sheet. If the film is only stretched in a single direction then that is termed the '1' direction. The two other orthogonal axes are labeled '2' and '3' as shown in Figure 5.10.

An annealing process at raised temperatures enhances the piezoelectric effect and the PVF2 material is then metalized to provide electrical connections to upper and lower faces of the film. The final processing stage involves applying an intense electric field of about 600 kV/cm at raised temperatures to polarize or orientate the molecules in the film, giving the desired piezoelectric properties. Electrically, the PVF2 polymer behaves like a capacitor. When pressure is applied to the film a charge, proportional to the pressure, appears on the surface of the material. The constant $d_{ab}$ relates pressure applied to the film in direction 'b' and the resulting charge appearing on face 'a' of the material. Because the top and bottom faces are metalized the charge generated in direction '3' is usually measured.

$$d_{ab} = \frac{charge\ /\ area}{force\ /\ area}\ C\ /\ N \tag{5.4}$$

Typical values for $d$ are (*Kynar Piezo Film Technical Manual* 1987):

$d_{31}$ = $23 \times 10^{-12}$ C/N;
$d_{32}$ = $3 \times 10^{-12}$ C/N; and
$d_{33}$ = $-32 \times 10^{-12}$ C/N.

If the output of a PVF2 sensor is fed into a high-impedance buffer amplifier (Figure 5.11(b))then the input voltage 'seen' by the amplifier is governed by the sum of the sensor, wiring, and amplifier input capacitances and the charge resulting from the applied force:

$$q = c\ .v_{in} \tag{5.5}$$

where:

$q$ = charge resulting from applied force (coulombs);
$c$ = sum of sensor and other parallel capacitance (farads); and
$v_{in}$ = amplifier input voltage (volts).

Thus, the buffer output voltage is reduced by the additional parallel capacitance of the wiring and amplifier input capacitance. The PVF2 polymer, wiring insulation, and buffer amplifier input represent a very large, but finite, parallel leakage resistance which causes the charge to leak away. Therefore, the output voltage due to a steady force decays away to zero over time and, for this reason, only changes in force can be detected by PVF2 sensors.

A charge amplifier can reduce the effects of stray capacitance and leakage resistance across the sensor. This scheme (Figure 5.11(c)) effectively transfers the sensor charge onto another capacitor $C$. The value of $C$ determines the output voltage for a given charge:

$$V_{out} = -\frac{q}{C} \tag{5.6}$$

Due to drift within the amplifier this circuit needs to be reset occasionally to prevent large errors building up over time. An alternative is to connect a resistor across $C$ to prevent charge build-up due to drift in the amplifier.

Dario and Buttazzo (1987) have developed a skin-like sensor based on PVF2 film. This sensor contains two force-sensing layers and has the additional capability of sensing thermal properties. Only the force-sensing properties will be described here. The sensor structure, illustrated in Figure 5.12, comprises a deep sensing layer ('dermal' sensor), a relatively thick, intermediate compliant layer, and a superficial thin sensing layer ('epidermal' sensor). The dermal sensor consists of a 5 x 7 array of sensor elements spaced 5 mm apart and is constructed by bonding PVF2 film to a pattern of electrodes. This layer is sensitive to normal force. The epidermal layer contains seven elements arranged in a hexagon (a circle of six with one in the center) at 5 mm spacing. These elements are highly sensitive to both deformation and temperature changes.

PVF2 polymer film is thin and flexible, allowing it to be easily molded around simple curves. Its response to applied loads is linear with a very large bandwidth. Unfortunately, the response does not extend down to DC and therefore steady loads cannot be measured directly. Essentially, the PVF2 material produces a charge output and this is prone to electrical interference and is difficult to multiplex in large sensor arrays. PVF2 is sensitive

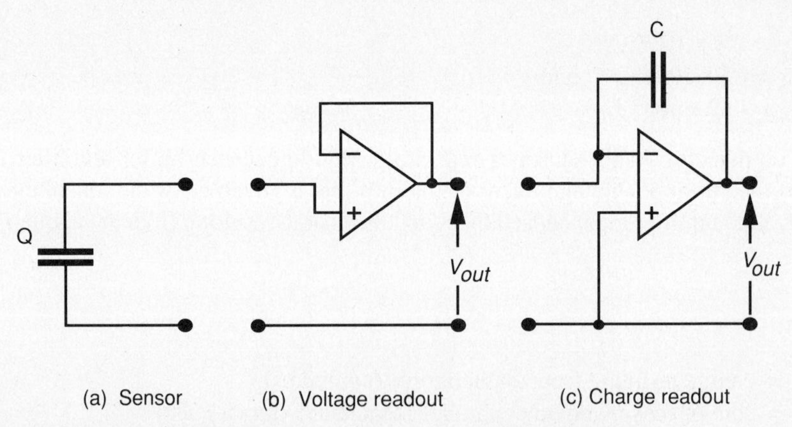


**Figure 5.11** Methods of reading the charge output of a PVF2 sensor

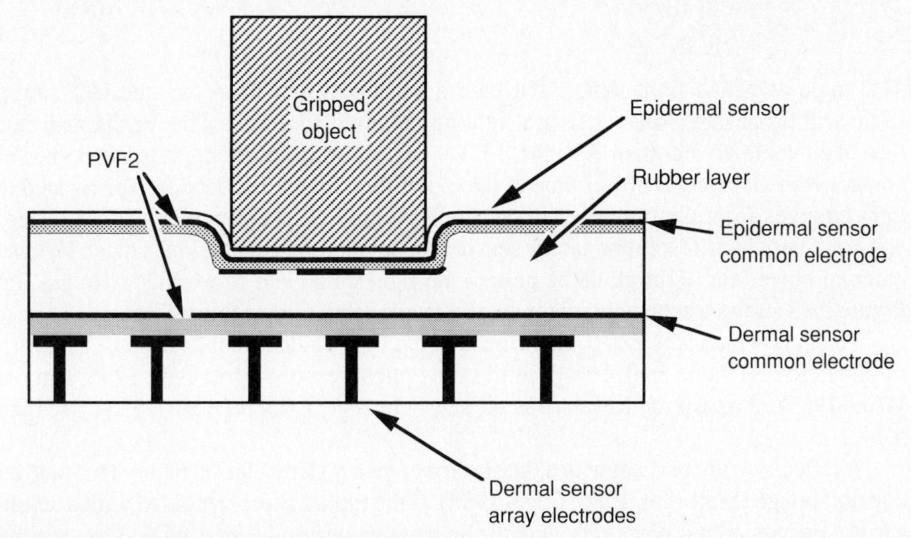


**Figure 5.12** PVF2-based tactile sensory skin
(From Dario et al. 1985, ©1985 IEEE)

to heat. Temperatures above 100°C destroy its piezoelectric properties. At lower temperatures PVF2 generates a voltage when its temperature changes (the pyroelectric effect). This effect can be useful for measuring temperature, but if a particular sensor can be affected by both load and temperature this may make it difficult to separate the two effects. There will be many applications for piezoelectric polymers in robotic sensing, especially in the detection of impacts and vibrations.

# 5.5 Optical

## 5.5.1 Frustrated internal reflection

A sheet of clear plastic can act as a light guide. Light introduced at one edge will propagate across the sheet by total internal reflection and emerge at the opposite edge. The property that determines the amount of light reflected from the interface between two dielectrics is their refractive index. Refractive index $n$ is the ratio of the velocity of light in a vacuum $c$ and the phase velocity of light in the dielectric medium $v$.

$$n = c / v \tag{5.7}$$

The refractive index of air is approximately 1 and that of perspex (polymethylmethacrilate) about 1.5. Perspex is a clear plastic suitable for making light guides. Total internal reflection occurs when light is propagating in the denser of two media and strikes the interface at less than the critical angle $\theta_c$ where:

$$\sin \theta_c = n_2 / n_1 \quad (\text{for } n_1 > n_2) \tag{5.8}$$

The angle at which light strikes the interface is measured from the interface normal. Light will propagate along a perspex light guide provided it strikes the perspex/air interface at an angle greater than $\theta_c$ ($\cong 41.8^\circ$). When the surface of the light guide comes into contact with an external object then at that point total internal reflection is frustrated and light emerges from the opposite side of the light guide. This principle has been used to construct very high-resolution tactile sensors. In practice, a flexible skin is placed between external object and light guide to protect the light guide and to exclude external light. Figure 5.13 shows a schematic diagram of a tactile sensor using this effect.

The light which emerges from the back of the light guide is detected either by an array of photodiodes (Tanie et al. 1986), solid state optical sensors (Mott et al. 1986; King and White 1985), or transported away from the sensor by optical fibers (Begej 1988). See Figure 5.14.

A reflective rubber sheet with a flat surface gives a high resolution binary (contact or no contact) image (Mott et al. 1986: Begej 1988). If the rubber sheet is molded with a textured surface then an output proportional to the area of contact, and hence applied force, can be obtained (Tanie et al. 1986; King and White 1985).

This kind of optical touch sensor can also be made sensitive to shear forces in the reflective rubber material. Shear forces are imaged by embedding specialized structures within the rubber skin material. Figures 5.15 and 5.16 show two examples.

The microlever proposed by Dario et al. (1988) distorts a rubber element when subjected to external forces. If the forces transmitted to the microlever contain shear components, the lever is pushed over and the patterned rubber element makes unequal contact with the light guide as shown in Figure 5.15.

An elastic sphere serves a similar purpose in a sensor proposed by King and White (1985). In this case shear forces displace the elastic sphere from its unloaded central position. The effect on the ball of normal and shear forces is shown in Figure 5.16. In both

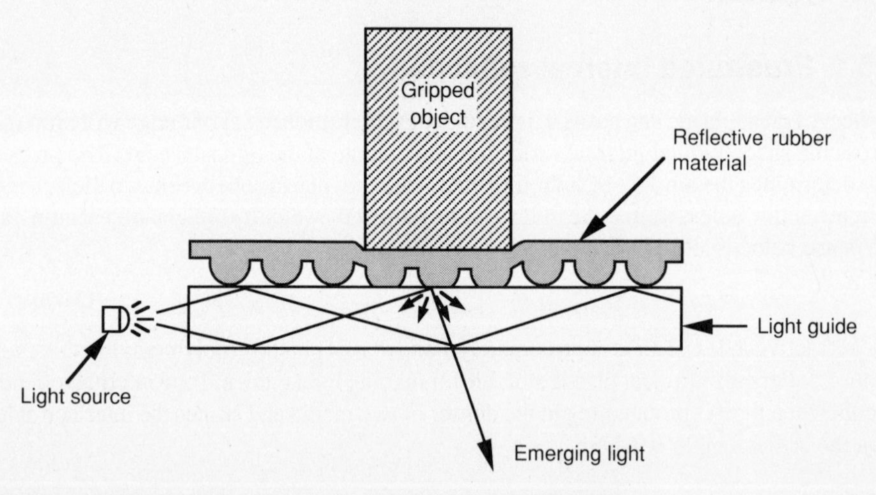


**Figure 5.13** Tactile sensor based on frustrated internal reflection

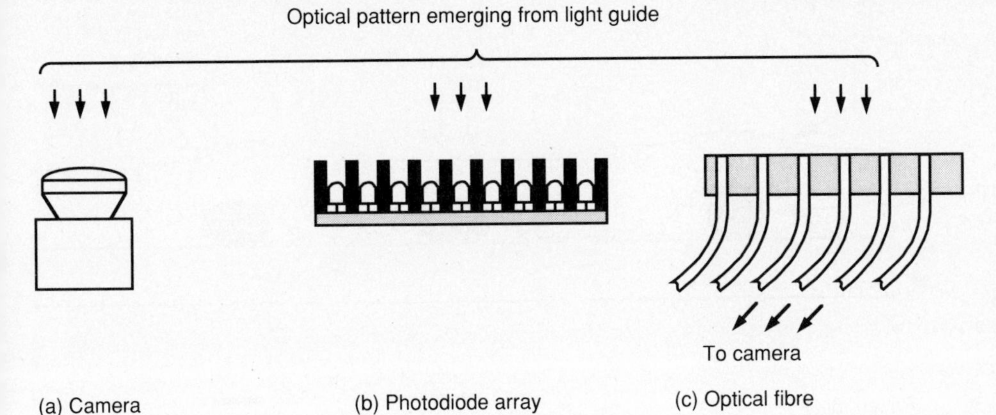


**Figure 5.14** Methods of detecting the emerging light

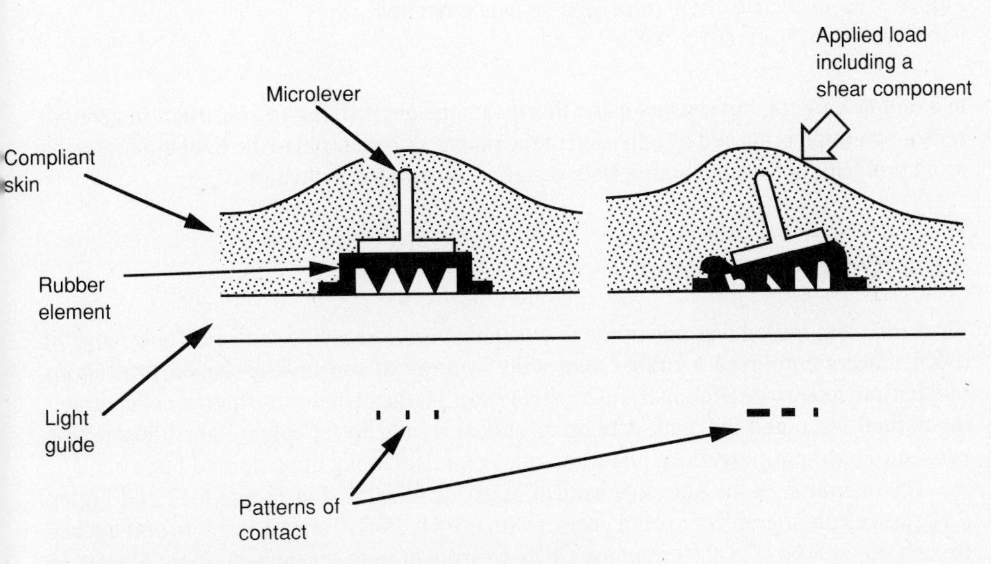


**Figure 5.15** Detecting shear forces with a microlever
(From Dario et al. 1988, redrawn with permission of the author)

of these cases the optical pattern created by distortion of the patterned rubber element and elastomer ball can be analyzed to gauge the amount of normal and shear forces present.

Sensors based on frustrated internal reflection can be molded to fingertip shape (Begej 1988) and are capable of forming very high-resolution tactile images, probably the highest resolution of any tactile sensor design. Because they modulate light, the transduction mechanism is immune to electromagnetic interference provided external light is excluded. Those touch sensors which use a lens to focus an image onto a CCD (charge coupled device) or DRAM (dynamic random access memory) optical sensor are too bulky to mount on the surface of most robot grippers or hands. Optical fibers have been used to transfer the image

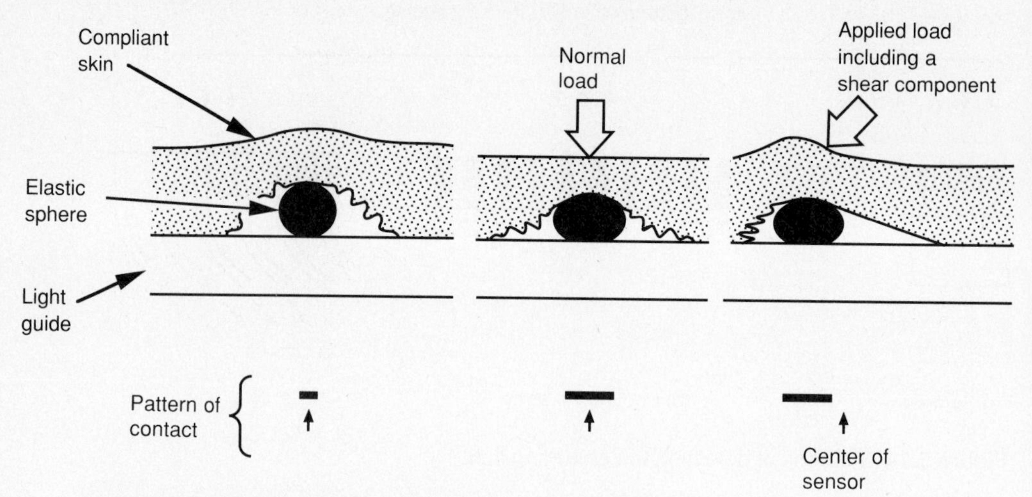


**Figure 5.16** Detecting shear force with an elastomer ball
(Redrawn from King and White 1985)

to a remote camera, but sensors using this technique are difficult to construct. In general, sensor operation is marred by adhesion of the rubber skin material to the light guide and the usual problems associated with rubber materials mentioned in Section 5.3.1.

## 5.5.2 Opto-mechanical

The Lord Corporation was one of the few manufacturers of tactile sensors. Their range of touch sensors employed a rubber skin with an array of mushroom-shaped projections molded into its surface (Rebman and Morris 1986). The head of the mushroom concentrates the normal force and the stalk acts as an optical shutter to modulate light transmission between a light-emitting diode and a photodetector depending upon normal force.

The operation of the opto-mechanical sensor is illustrated in Figure 5.17 and Figure 5.18 shows a photograph of a robot gripper with Lord LTS-210 array force sensors attached to each finger. The LTS-210 contains a 10 x 16 array of opto-mechanical force sensors on a 1.8 mm horizontal and vertical spacing. Six axis force sensing is also incorporated in the LTS-210.

Construction of the Lord Corporation touch sensors is quite labor intensive. Each sensor site contains a photoemitter and photodetector. These are individually matched and trimmed with an associated resistor to equalize their responses. For closer matching, the characteristics of each sensor site is recorded in a look-up table and used to apply additional compensation in software. Because the deflecting element is made of rubber, the sensor response will be subject to the usual problems of creep, hysteresis, memory, and temperature variation. There seems to be no fundamental problem to prevent construction of a fingertip-shaped sensor using this design.

Elastomers and rubbers have relatively poor mechanical properties when used as the deformable member in a load cell. An opto-mechanical touch sensor has been reported by

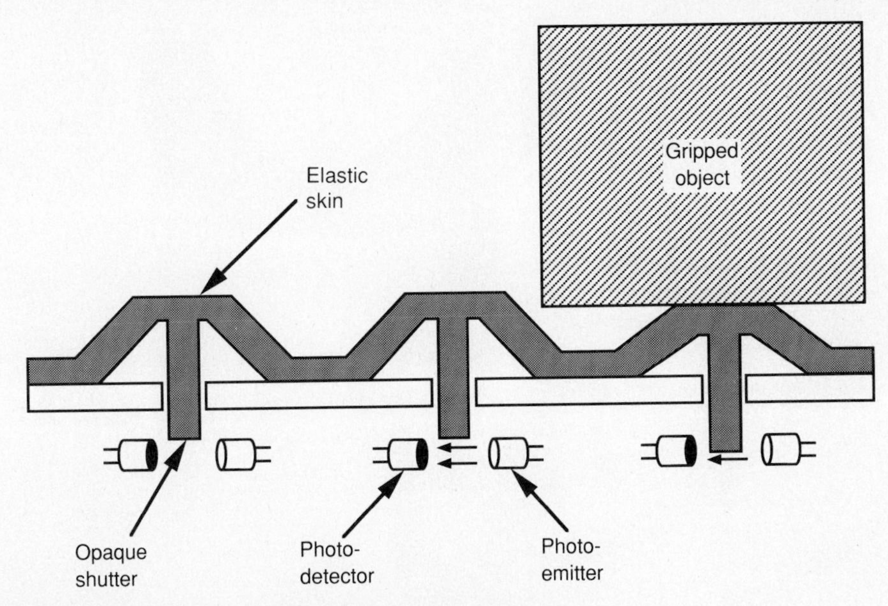


**Figure 5.17** An opto-mechanical array touch sensor
(Redrawn from Rebman and Morris 1986, courtesy of IFS Publications)

Maalej and Webster (1988) which uses a U-shaped beam of spring steel as the deformable member. Other spring materials such as phosphor-bronze and beryllium-copper can also be used. (*Note*: beryllium is very toxic.) A diagram of the sensor is shown in Figure 5.19. This design has low hysteresis, good repeatability and reasonably linear output. Large size (4 mm x 6 mm x 3 mm) limits the spatial resolution that can be achieved using this kind of sensor.

## 5.5.3 Fiber-optic

Optical fibers are used as light guides to carry optically coded information and light energy with the same convenience that electrical signals and energy are carried by electrical wiring. However, optical fibers can also form part of the transducer mechanism, as well as transferring light energy to and data from a touch sensor.

### *5.5.3.1 Light coupling between adjacent fibers*

Light propagates along an optical fiber with very little lost due to radiation; however, if the surface of the fiber is roughened then at that point light can leave and enter the fiber. If two optical fibers pass close to each other and both have a roughened surface then light can pass between the fibers. Light coupling between adjacent optical fibers is a function of their separation. The sensor design shown in Figure 5.20 uses D-section cords made of closed-cell wetsuit type rubber as a deformable member and the light coupling between crossed plastic optical fibers to measure the resulting deflection (Schoenwald et al. 1987). Loads applied normal to the sensor surface compress the D-section elastomer cords moving the fibers closer together and thus increase the light coupling.

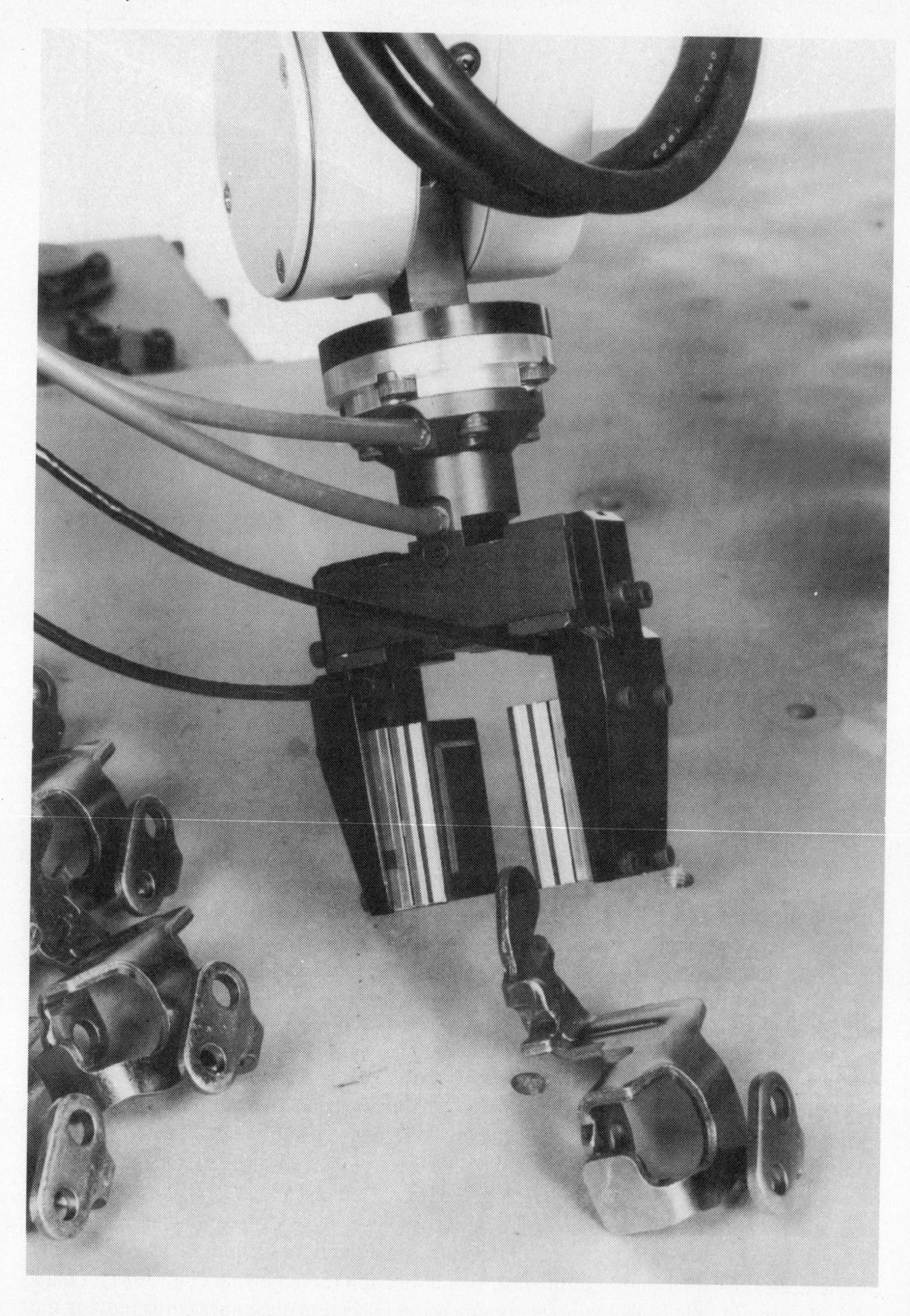

**Figure 5.18** The Lord LTS-210 tactile sensor mounted on a robot gripper
(Photograph courtesy of Lord Corporation, Industrial Automation Division, Cary, North Carolina)

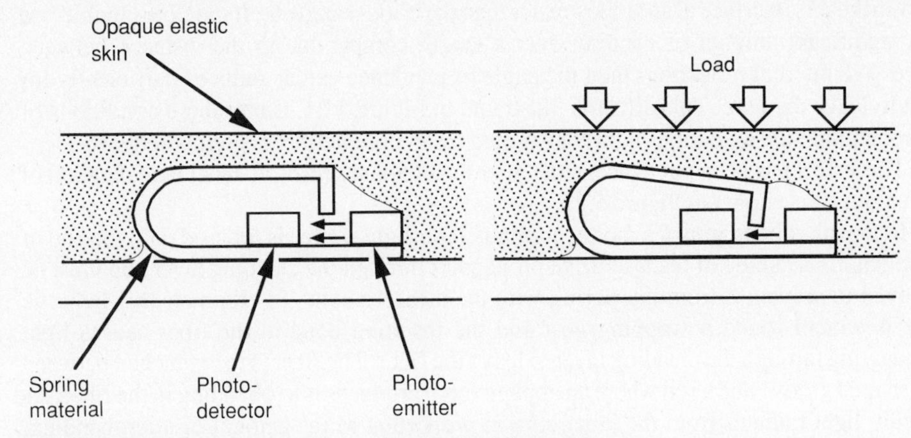


**Figure 5.19** Metal spring opto-mechanical touch sensor
(Redrawn from Maalej and Webster 1988, ©1988 IEEE)

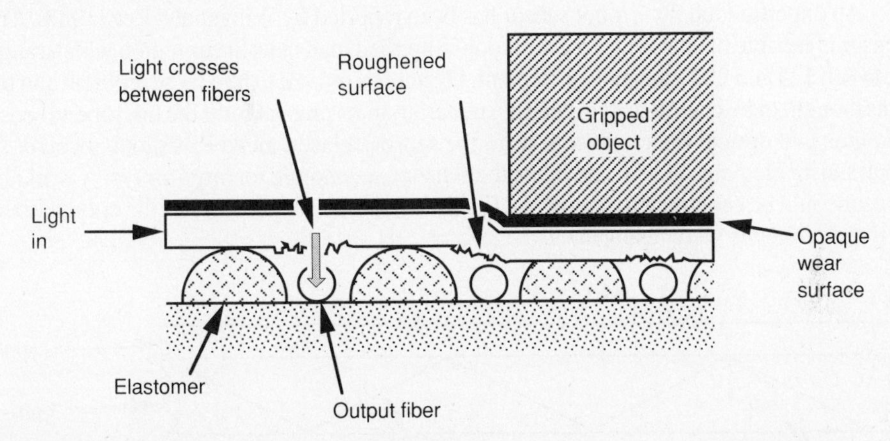


**Figure 5.20** Optical fiber sensor based on varying coupling between crossed fibers
(Redrawn from Schoenwald et al. 1987, ©1987 IEEE)

The sensor consists of parallel input fibers laid at right angles across a group of parallel output fibers. To read the fiber separation at one cross point, light is injected into the associated input fiber and the resulting light coupled to the corresponding output fiber is measured. A 4 x 4 array with 1 cm spacing has been reported by Schoenwald et al. (1987). Parasitic light paths will be present in this design. Unfortunately, the guarding and virtual earth techniques for overcoming parasitic paths (see Chapter 4) do not work for light. This sensor is flexible and can conform to complex curved surfaces. It has the advantage of noise immunity associated with optical transducers and the imperfections introduced by the use of elastomer materials as the deformable member.

### *5.5.3.2 Bending losses in optical fibers*

Light propagates through an optical fiber by repeated internal reflection from the core/cladding interface. As explained in Section 5.5.1, for total internal reflection to occur, light must strike this interface at an angle greater than the critical angle $\theta_c$. If the fiber is subjected to a significant amount of bending over a length comparable to the distance between successive internal reflections then the angle of incidence can be reduced sufficiently for light to leave the core. This effect is illustrated in Figure 5.21. Under these conditions of microbending, the amount of light transmitted by an optical fiber is greatly reduced.

This effect has been put to use in the microbend touch sensor. A schematic diagram of such a sensor is shown in Figure 5.22.

In the microbend sensor a monochromatic laser light source is focused onto the end of an optical fiber. Some of the laser light propagates through the cladding layer and must be removed to prevent it from interfering with the measurement. To eliminate this light the fiber is wound round a stripping post and the resulting bend in the fiber causes light propagating through the cladding layer to leave the fiber. The fiber is then pinched between a V-shaped groove and a rod where an applied force produces microbending in the fiber. As a result, light radiates from the inner core in proportion to the amount of microbending. Some of the emerging light propagates inside the cladding layer and must be removed by another stripping post before the light remaining in the inner core is detected by a PIN diode.

An experimental 2 x 2 robot sensor has been reported by Winger and Lee (1988). This sensor is capable of detecting a 5 g variation in applied load in its linear region which ranges between 125 and 225 g per sensor element. Hysteresis proved to be a large problem and this was thought to be caused by the cladding material. In its present form the microbend sensor requires two optical fiber connections to the sensor, a laser, and a PIN diode receiver for each sensor element. This would make the sensor uneconomic for large arrays. A workable sensor could be constructed, having the flexibility and noise immunity of the optical fibers, if these problems were resolved.

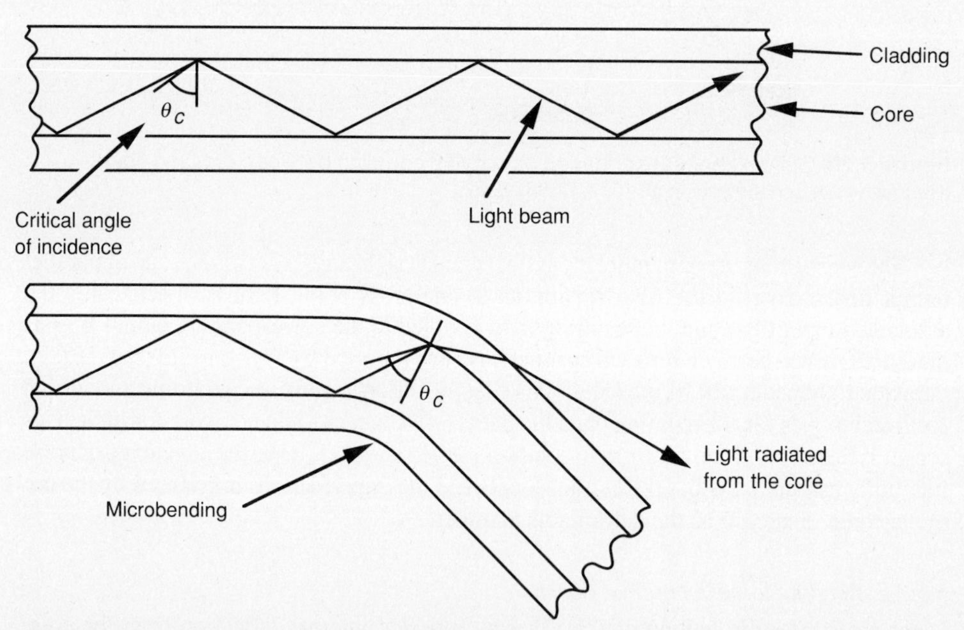


**Figure 5.21** Light radiation due to microbending

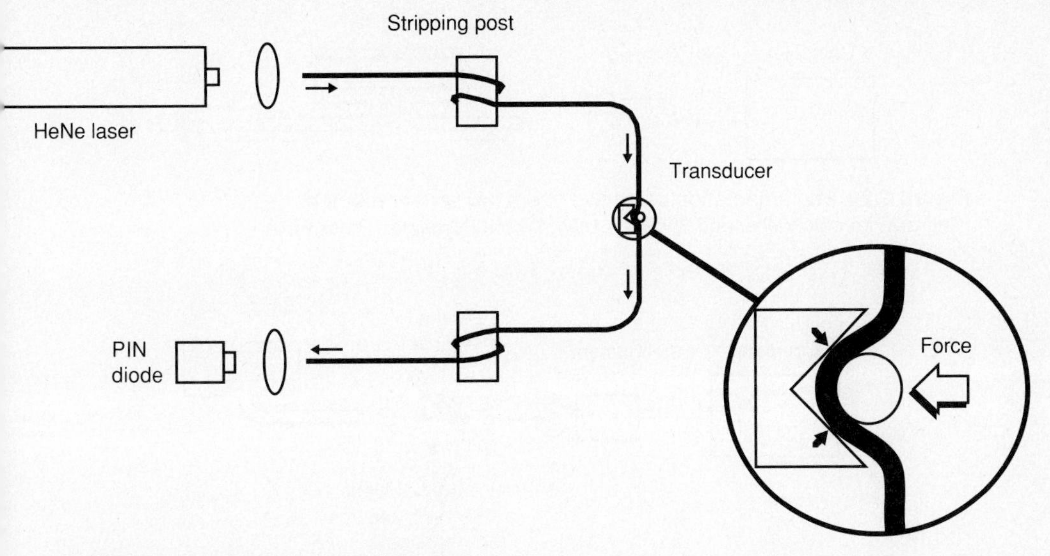


**Figure 5.22** The fiber microbend sensor
(Redrawn from Winger and Lee 1988, ©1988 IEEE)

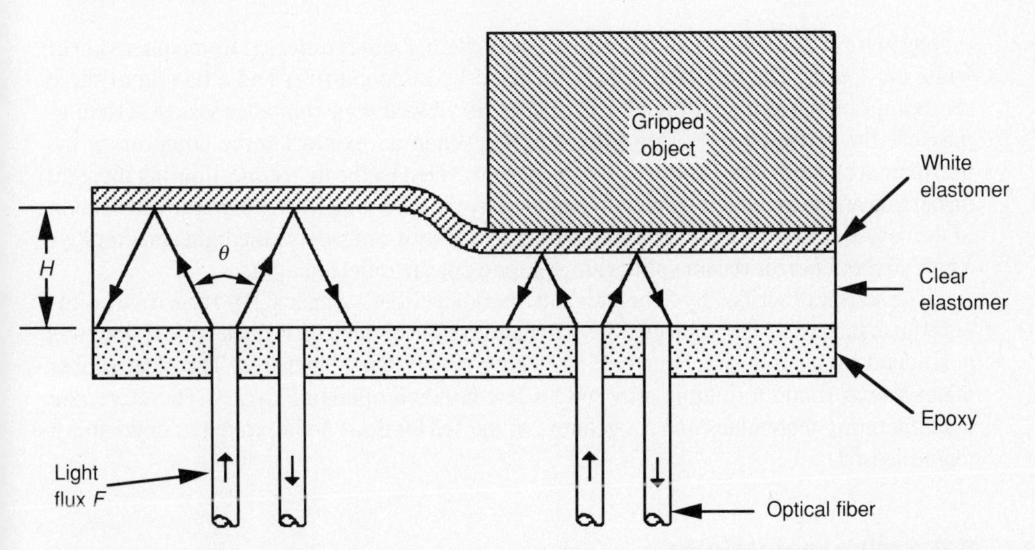


**Figure 5.23** Tactile sensor based on a deformable elastic reflective surface and fiber-optic technology
(Redrawn from Schneiter and Sheridan 1984, ©1984 Pergamon Press PLC)

*5.5.3.3 Optical skin thickness sensor*

This sensor (Schneiter and Sheridan 1984) determines the thickness of a transparent, deformable elastomer layer by measuring the intensity of light reflected back from the far side of the layer. Figure 5.23 shows a cross-section of the sensor.

Light is introduced into the sensor via an optical fiber. A widening cone of light

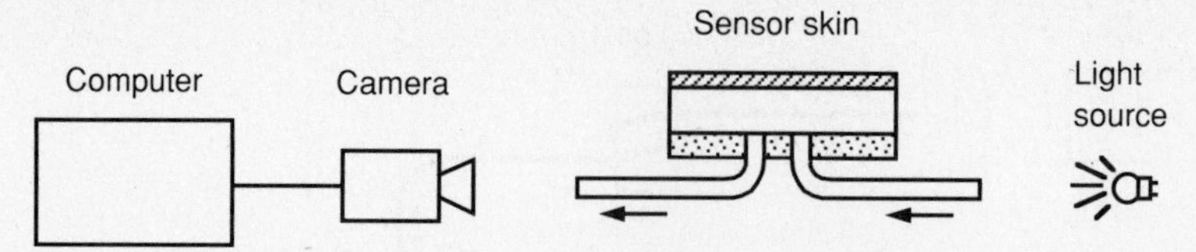


**Figure 5.24** Implementation using two fibers per sensor element
(Redrawn from Schneiter and Sheridan 1984, ©1984 Pergamon Press PLC)

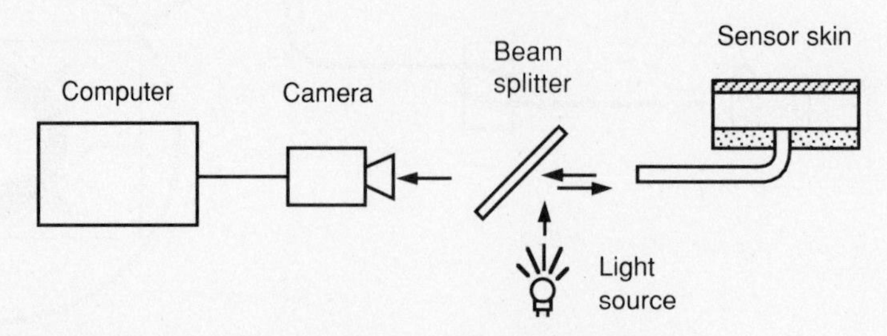


**Figure 5.25** Implementation using one fiber per sensor element
(Redrawn from Schneiter and Sheridan 1984, ©1984 Pergamon Press PLC)

propagates out through a layer of transparent elastomer and is reflected by an outer skin of white elastomer. The reflected light is received by a second fiber and a bundle of these receiving fibers, from an array of sensor sites, is viewed by a computer vision system to measure the reflected light (see Figure 5.24). When an external force compresses the transparent elastomer this shortens the distance traveled by the light cone, limiting the light dispersion and thus reducing the light gathered by the receiving fiber. An improved version of the sensor (Figure 5.25) uses a single fiber to transmit and receive the light and employs a half silvered mirror (beam splitter) to separate out the reflected light.

The sensor described by Schneiter and Sheridan (1984) contains 2100 sensitive points per square inch (6.45 $cm^2$) and exhibits a dynamic range of only 18:1. Each optical fiber was positioned by hand during construction, which is very labor intensive. The clear rubber material was found to fatigue after only a few hundred operating cycles. Therefore, the manufacturing techniques and ruggedness of the sensor must be improved to make this a viable sensor.

## 5.4.4 Photoelasticity

Photoelasticity can be used to measure stresses in a sample of optically active material. Consider the experimental set-up shown in Figure 5.26.

(a) Light radiated from a light source contains many waves of differing polarization and amplitude. Both polarization and amplitude vary with time. (Light exhibits a transverse wave motion in its direction of propagation and polarization represents the orientation of vibration of this wave.)

(b) Polarizer 1 only allows through components of each wave having a particular plane of polarization, and blocks those components which are at right angles.

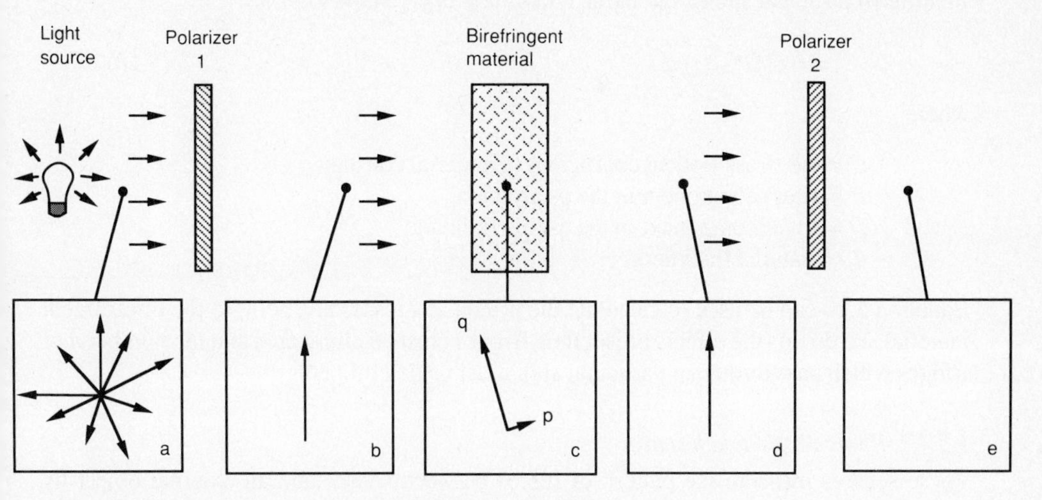


**Figure 5.26** Measuring stresses using photoelasticity

(c) Upon entering the birefringent material light is split into two components, p and q, polarized at right angles. These components are aligned with the principal stresses in the birefringent material. (Principal stresses are planes of maximum and minimum stress in the material.)
(d) If the two waves emerge from the birefringent material with the same relative phase that they had when they entered then the original wave is reconstructed.
(e) Polarizer 2 is rotated 90° with respect to Polarizer 1 and therefore blocks transmission of the reconstructed wave.

If the two waves do not emerge with the same relative phase that they had when they entered the birefringent material then an elliptically polarized wave results. Part of this elliptically polarized wave is passed by polarizer 2. Dark areas, or 'fringes', where light is not transmitted through the system, are the result of two effects:

1. Isoclinics — areas where light from polarizer 1 is in line with one of the principal stress axes. In this case the light is not split into two components and therefore emerges unaltered. This effect provides information about the orientation of principal stresses in the birefringent material.
2. Isochromatics — result when light is split by the birefringent material but emerges with the same phase that it entered. This will be true if the principal stresses in the material are identical or differ by an amount which produces an integral number of phase rotations. Isochromatics can be represented mathematically by :

$$r = n.\lambda \tag{5.9}$$

where:

$n$ = an integer;
$\lambda$ = the wavelength of light; and
$r$ = relative retardation.

In terms of principal stresses, relative retardation is given by:

$$r = C\,(P - Q)\,d \qquad (5.10)$$

where:

$C$ = the stress-optical coefficient (a material constant);
$P$ = stress component in the p-direction;
$Q$ = stress component in the q-direction; and
$d$ = material thickness.

Equation 5.10 can be used to calculate the principal stresses at a point in the birefringent material. To do this the effects of isoclinic fringes must be eliminated and the number $n$ of fringes which pass through a particular spot must be determined.

*5.5.4.1 Photoelastic touch sensor*

These sensors measure the pattern of forces between sensor and an external object by imaging the resulting stresses induced in a layer of photoelastic material. A sensor has been proposed by Cameron et al. which measures the forces applied to a sheet of birefringent material (Cameron et al. 1988). As shown in Figure 5.27 the sheet of birefringent material would be illuminated by circularly polarized light to visualize isochromatic fringes. A CCD camera records the resulting stress pattern. It has been suggested that the camera image can

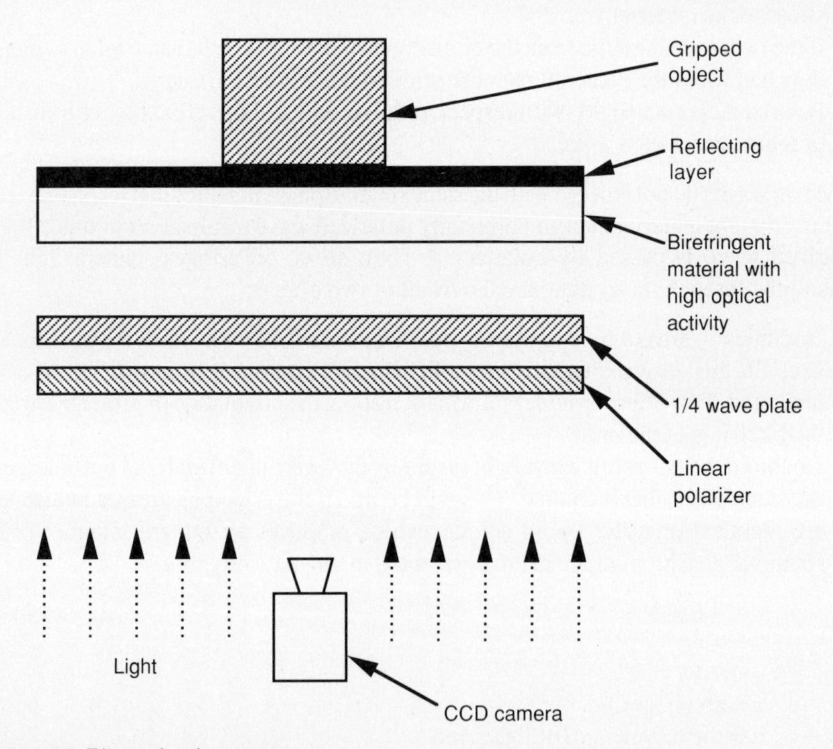


**Figure 5.27** Photoelastic sensor

be analyzed to determine the distribution of stresses within the birefringent plastic. Deconvolving the stress distribution with the point load function then produces the force profile caused by the indenting object. A similar sensor has been proposed by Eghtedari and Morgan (1989) as a means of measuring slip. Their sensor employs plain polarizers and therefore both isochromatic and isoclinic fringes are imaged. They did not have to analyze this more complicated fringe pattern because they were only looking for changes in the image which indicate the slipping of a gripped object.

This kind of sensor should give good results in terms of linearity, hysteresis, creep, and memory because the effect has been successfully applied to photoelasticity measurements for many years. The optical system behind the sensing surface of the sensor will make it bulky and difficult to accommodate within a robot hand or gripper. There would also be complications involved in molding the sensor to fingertip shape. Protection must be arranged for the birefringent sheet if it is to survive abrasion and impacts.

## 5.6 Magnetic

### 5.6.1 The Hall effect and magnetoresistance

The Hall effect is closely related to the motor effect observed as a force on a current-carrying conductor in a magnetic field (where the current direction and orientation of magnetic field are at right angles). The motor force acts directly on the charge particles (electrons in the case of metal wires) making up the current and the force is transferred to the current-carrying conductors. In Figure 5.28, if current $J_z$ flows through a conductive material in the *z*-direction and a magnetic flux $B_y$ is established in the *y*-direction, then charge carriers will experience a force in the *x*-direction. Charge will tend to accumulate on the bottom surface of the sample, producing a resulting Hall potential $E_H$.

Current $J_z$ flowing through the sample is given by:

$$J_z = N.q.v_d \tag{5.11}$$

where:

$N$ = number of charge carriers per unit volume;
$q$ = the charge on each charge carrier (coulombs); and
$v_d$ = average carrier drift velocity (m/s).

The velocity of charge carriers depends upon the applied electric field and charge carrier mobility:

$$v_d = m.E_z \tag{5.12}$$

where:

$m$ = charge carrier mobility ($m^2$/v.s); and
$E_z$ = electric field in the *z*-direction (v/m).

The force acting on a charge moving in a combined electric and magnetic field is:

$$\boldsymbol{F} = q\,(\boldsymbol{E} + \boldsymbol{v} \times \boldsymbol{B}) \tag{5.13}$$

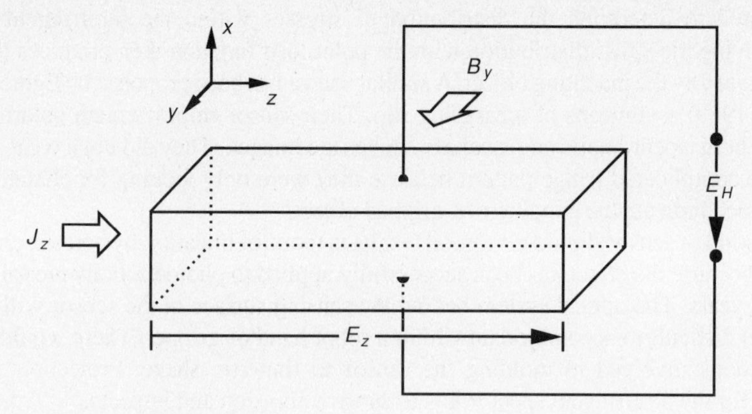


**Figure 5.28** The Hall effect voltage

where **x** is the vector cross-product. In the steady state, force produced by interaction of the moving charge and magnetic field is balanced by the force on the charge carrier caused by the Hall voltage electric field. A charge particle traveling with average velocity passes through undeflected. This balance of forces is represented by the following equation:

$$q.E_H = -q.\, m.E_z \times B_y \tag{5.14}$$

Therefore:

$$E_H = -m.E_z.\, B_y.i_x \tag{5.15}$$

where $i_x$ is a unit vector in the $x$-direction. Equation 5.15 can be rewritten as:

$$E_H = J_z.\, B_y.\, R \tag{5.16}$$

which shows that the Hall voltage is directly proportional to current $J_z$ flowing through the sensor and applied magnetic field $B_y$. Hall coefficient $R$ is given by:

$$R = \frac{1}{N.q} = \frac{m}{s} \tag{5.17}$$

where conductivity $s$ is:

$$s = N.q.m \tag{5.18}$$

It can be seen that $R$ is inversely proportional to carrier concentration and this is the reason why the Hall effect is more pronounced in semiconductor materials.

It has already been noted that a charge carrier traveling at the average velocity will pass through the Hall sensor without being deflected. However, charge carriers have a range of velocities and those that are traveling at a velocity other than the average are deflected and take a longer path to travel the length of the sensor. Effectively the deflected particles have a lower mobility and this shows as an increased electrical resistance. Charge particle deflection and hence sensor resistance is affected by the applied magnetic field. This is the magnetoresistive effect.

Note that, for a particular configuration of electrodes, a Hall effect sensor is only sensitive to magnetic fields in one direction (this would be the *y*-direction in Figure 5.28). By contrast, the magnetoresistive effect can be used to detect magnetic fields having any orientation within a plane (in Figure 5.28 this would be the *x–y* plane).

Both the Hall effect and magnetoresistance can be used to measure magnetic field intensity. For touch sensor designs, the magnetoresistive effect has advantages in terms of simplicity (a magnetoresistive sensor requires two connections rather than four for a Hall sensor) and sensing orientation (a magnetoresistive sensor can detect a magnetic field in a plane perpendicular to the direction of current flow, while a Hall sensor only detects a field mutually perpendicular to the direction of current flow and the orientation of the electrodes detecting the Hall voltage).

#### *5.6.1.1 A torque-sensitive tactile array*

In this sensor design, a sheet of silicone elastomer contains an array of embedded magnetic dipoles (Hackwood et al. 1983). Beneath each dipole four Permalloy magnetoresistive sensors are mounted on a rigid substrate. Permalloy is an alloy of 19 per cent iron and 81 per cent nickel which exhibits a strong magnetoresistive effect. Magnetic field strength at each of the four magnetoresistive sensors is used to determine both position and orientation of the magnetic dipole. Normal forces, tangential forces, and torques are transmitted through the rubber and can be determined from the change in height and orientation of each magnetic dipole.

An experimental sensor containing two elements mounted 1.6 mm apart is reported to have worked well with no detectable hysteresis. Magnetic screening would be required to protect the magnetoresistive sensors from external magnetic fields. Because silicone rubber is used as the deformable member in this sensor, the characteristics of the rubber would tend to limit the sensor performance in terms of creep, linearity, memory, and sensitivity to temperature.

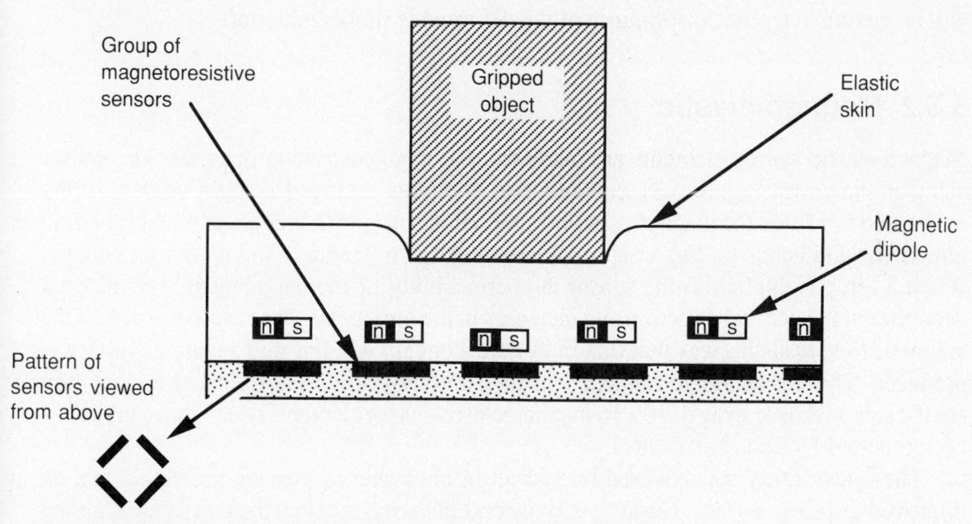


**Figure 5.29** Magnetoresistive sensor using magnetic dipoles
(Redrawn from Hackwood et al. 1983, ©1983 MIT Press)

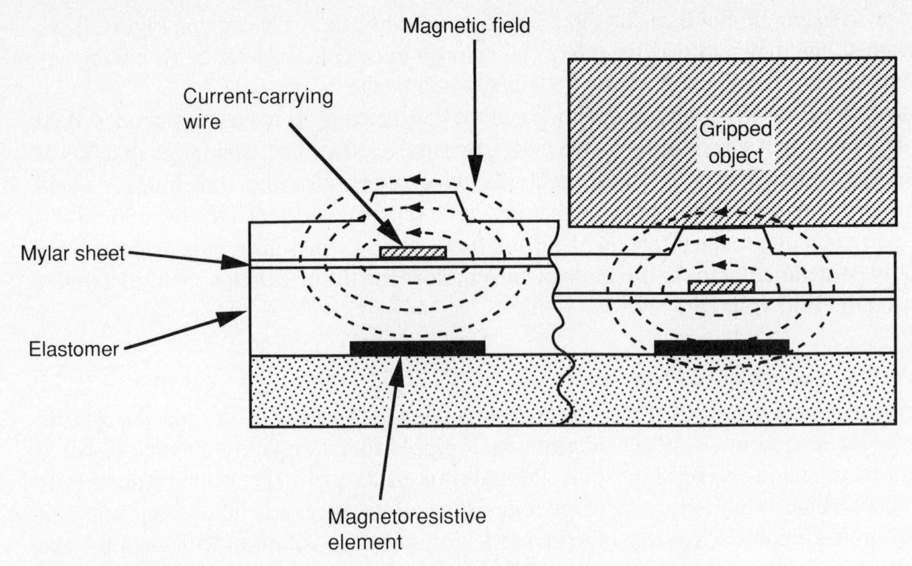


**Figure 5.30** Magnetoresistive sensor using current-carrying wires
(Redrawn from Vranish 1986, courtesy of IFS Publications)

*5.6.1.2 Magnetoresistive skin*

An alternative form of magnetoresistive sensor has been proposed by Vranish (1986). In this sensor (Figure 5.30), current-carrying wires provide the magnetic field. Row and column addressing can be organized by using the current-carrying wires to select rows of sensor elements and connecting magnetoresistive elements in series to produce columns. Vranish suggests that a sensor spacing of 2.5 mm should be possible with this sensor. Once again magnetic screening will be of great importance for this sensor and the sensor characteristics will be governed by the performance of the deformable rubber material.

## 5.6.2 Magnetoelastic

Magnetoelastic sensor elements are made from a magnetostrictive material. The sensor element shown on Figure 5.31 contains two windings arranged at right angles. In the unstressed condition the magnetostrictive material is isotropic (has equal permeability in all directions) and hence no flux coupling between the two windings and no output voltage. When a force is applied to the sensor the permeability of the magnetostrictive material decreases in the vertical direction and increases in the horizontal. The result is to distort the magnetic field in such a way that flux links the secondary winding and an output voltage is produced. These sensors are rugged, sensitive, and have low hysteresis and temperature coefficient. A sensor array of 16 x 16 magnetoelastic sensor elements on 2.5 mm centers has been reported by Luo et al. (1984).

The sensor array was covered by a sheet of elastomer to provide protection and an improved gripping surface. Good sensitivity and linearity, and low hysteresis are claimed for the sensor. The sensors and their associated circuits are relatively complicated. If the sensor elements can be reduced in size and screened from external magnetic fields they

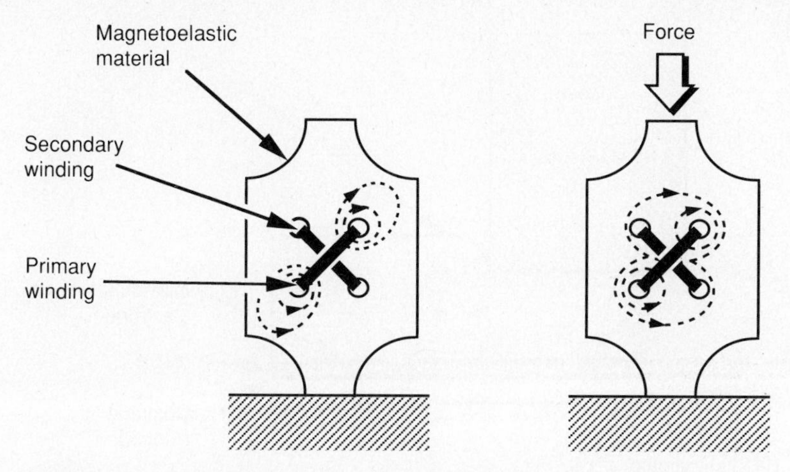


**Figure 5.31** Changes in flux distribution caused by applied force
(Redrawn from Luo et al. 1984)

could form a useful touch sensor. Magnetic screening can be arranged by enclosing the screened area within a layer of material having high magnetic permeability. In the case of a tactile sensor, the screening layer must also be flexible.

## 5.7 Ultrasonic

Ultrasonic thickness gauges have been in use for many years to measure the thickness of paint layers, metal sheets, etc. We can measure the time taken for an ultrasonic pulse to travel through the material, reflect off the back surface and return. If the speed of propagation of the ultrasonic wave in the material is known then the material thickness can be calculated. This principle has been used to construct a tactile sensor by using ultrasonic transducers to measure the thickness of a flexible elastomer layer at many closely spaced points. Referring to Figure 5.32, if the elastic skin has a thickness $H$ and the velocity of sound in the elastic skin material is $v_s$, then the time $t_f$ for an ultrasonic pulse to travel from the transducer to the surface of the skin and return is:

$$t_f = \frac{2H}{V_s}$$

*Example:* If the speed of sound in the elastic skin material is $10^3$ m/s and skin thickness is 3 mm, then the delay will be 6 μs.

As an object indents into the flexible layer the thickness of the layer is reduced and measuring $H$ at many closely spaced points provides an image of the indentation. Grahn and Astle (1984) claim a dynamic range of 2000:1 and a spatial resolution of 0.5 mm for this sensor.

This sensor could certainly be molded to fit the shape of a fingertip. The circuits to

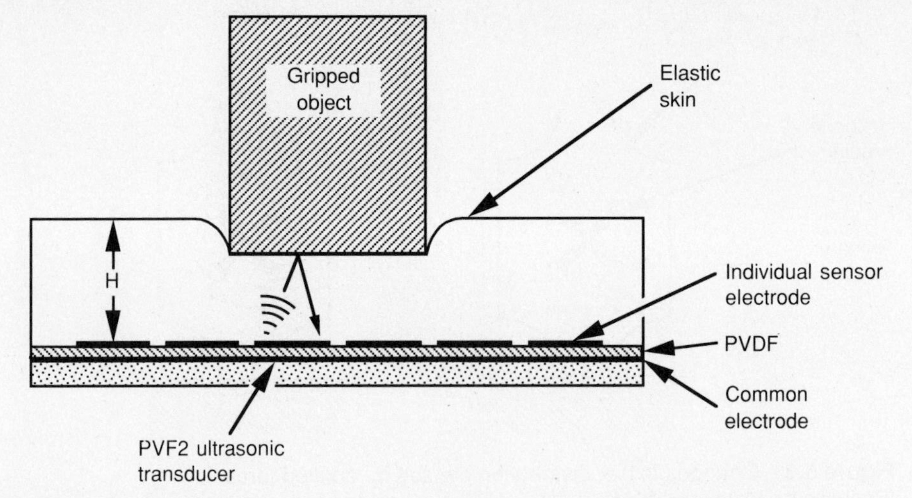


**Figure 5.32** Tactile sensor using ultrasonic pulses to detect elastic skin thickness
(Redrawn from Grahn and Astle 1986, courtesy of Cambridge University Press and the author)

generate and detect ultrasonic pulses from many closely spaced sites must be relatively complicated. If the sensor were immersed in a liquid or touched an object with a similar acoustic impedance to the skin material it is possible that the ultrasonic pulse would propagate past the surface of the skin material without being reflected.

## 5.8 Capacitive

Over small distances capacitance can be used to measure the separation between two conductive plates. Ignoring fringing fields at the edge of parallel capacitor plates the capacitance between the plates is:

$$c = \frac{\varepsilon_0 \cdot \varepsilon_r \cdot A}{s} \tag{5.20}$$

where:

$A$ = capacitor plate area ($m^2$);
$c$ = capacitance between the plates (farads);
$\varepsilon_0$ = absolute permittivity of free space ($8.854 \times 10^{-12}$ F/m);
$\varepsilon_r$ = relative permittivity of the dielectric between the capacitor plates; and
$s$ = separation between the capacitor plates (m).

Shear forces can alter the area $A$ of overlap between two plates and normal force can affect the plate separation $s$. Therefore, in principle, capacitance could be used to measure both shear and normal forces. However, careful mechanical design is required to prevent a sensor reacting to both shear and normal forces at the same time which would make it difficult to separate the two effects.

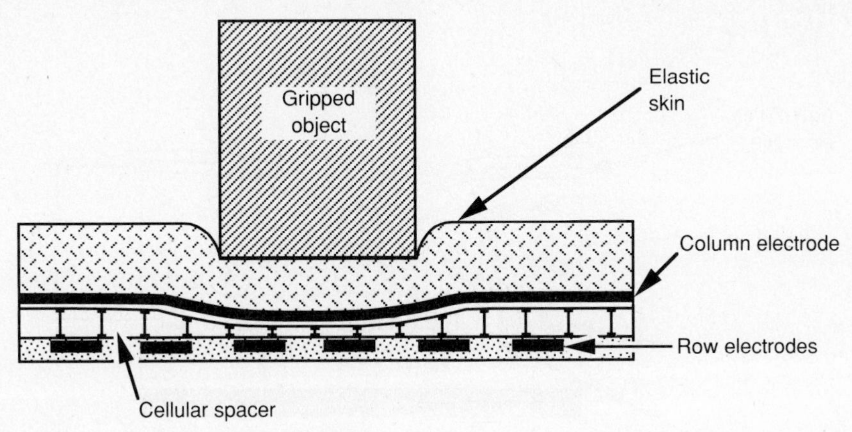


**Figure 5.33** Capacitative touch sensor

Siegel et al. (1985) have described a tactile sensor (see Figure 5.33) which uses an injection-molded silicone rubber honeycomb material to form the dielectric between upper and lower electrodes. The cellular nature of the dielectric allows it to compress as external forces are applied to the sensor surface. An outer conducting elastic skin physically protects the capacitors and screens them from outside electric fields. The upper column electrodes are plated on a Mylar sheet. Beneath the elastic/dielectric layer, row electrodes are etched from printed circuit material. A multiplexing scheme allows the capacitance at the cross point of any row and column electrode to be measured and hence the deflection at that point is determined. The sensor consists of an 8 x 8 array of sensor points with 1.9 mm spacing formed into a cylinder. The resulting sensitive fingertip is designed to be mounted on the Utah-MIT dextrous hand (the Utah-MIT dextrous hand is described in Chapter 8).

Capacitative sensor arrays can be molded to conform to compound curves and provide very accurate measurements of skin deflection. The properties of the sensor, in terms of hysteresis, creep, memory, non-linearity, etc., are governed by the deformable elastic material between the capacitor plates.

## 5.9 Electrochemical sensor

Chemical-impregnated gels have been formulated to make them sensitive to deformation. De Rossi et al. (1988) have reported an electrochemical sensor based on the phenomenon of streaming potentials. Figure 5.34 shows a cross-sectional view of the sensor. An ionized gel disk 1 cm in diameter and 0.4–0.5 mm thick is made from polyacrylic acid and polyvinylic acid. The gel contains an immobile negative charge which is balanced by a mobile positive charge. When pressure is applied to the gel, positively charged liquid is forced out of the gel and an inhomogeneity of charge is formed which constitutes the streaming potential. Silver/silver chloride electrodes pick up this potential. The sensor can detect low-frequency deformations but has no steady state response. Further development is required to miniaturize the sensor and improve its sensitivity.

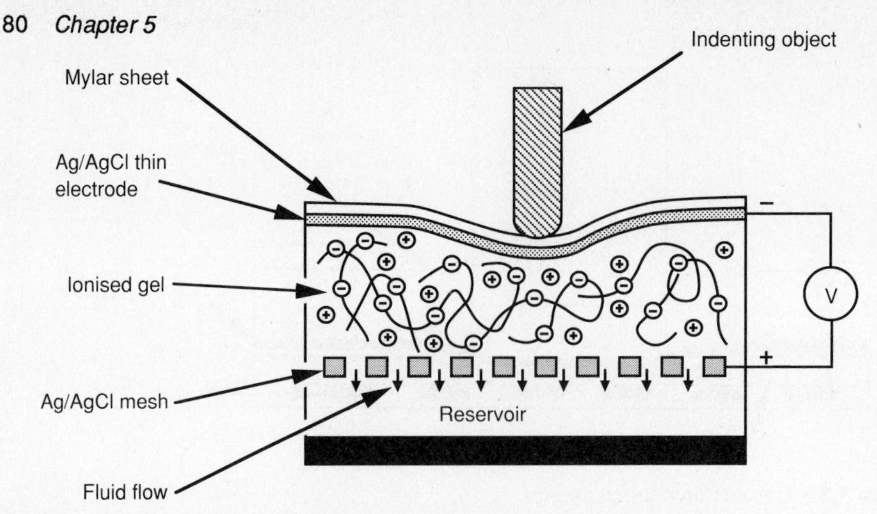


**Figure 5.34** Schematic diagram of the streaming potential sensor (Redrawn from De Rossi et al. 1988, ©1988 IEEE)

A touch sensor has been proposed which uses polyelectrolytic gels to detect applied forces. The application of a force causes migration of different types of mobile ions within the gel. Ion-selective electrodes can be used to measure the concentrations of these ions which are a unique function of the distortion of the polyelectrolytic gel. Thus, in principle, the ion concentrations can be used to determine stresses applied to the sensor with a frequency response which extends down to the steady state (De Rossi et al. 1989). The gel component of electrochemical sensors can be formulated to give the sensor surface similar compliance characteristics to the human skin.

## 5.10 Conclusion

Plots of typical touch sensor images are shown in Figures 5.35 and 5.36.

The majority of sensors considered in this chapter are based upon the principle of measuring the deformation of an elastomer layer. Sensor performance, in terms of hysteresis, creep, memory, non-linearity, and temperature effects are limited by the characteristics of the elastic material. Elastomeric materials are essentially incompressible and therefore radial expansion is limited by surrounding material. It takes a considerable force to indent an object into an elastomer layer if the area of contact is large.

The sensors considered in this chapter all exhibit relatively little compliance and are mainly suited to imaging relatively flat objects. As Brady (1988) has observed, they are good at determining object position and an approximation of surface normals, but they are poor at measuring surface curvature. In the next chapter we will consider a group of sensors which have a large compliance which enables them to determine surface contours.

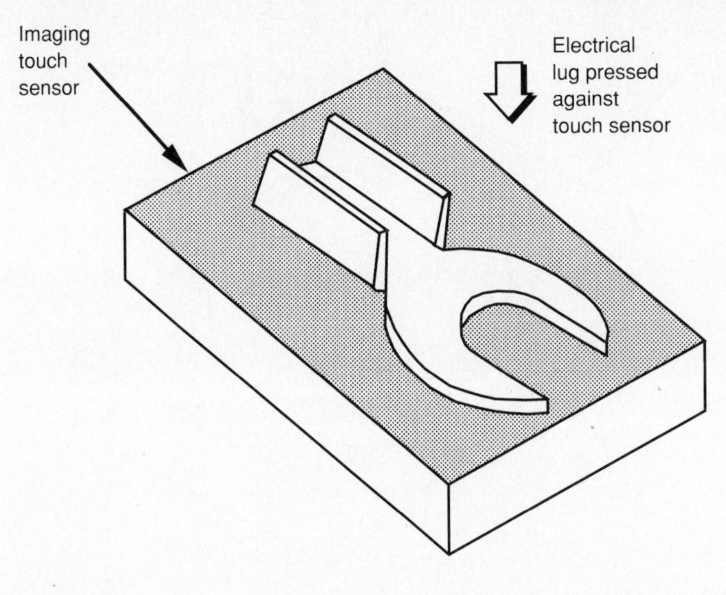


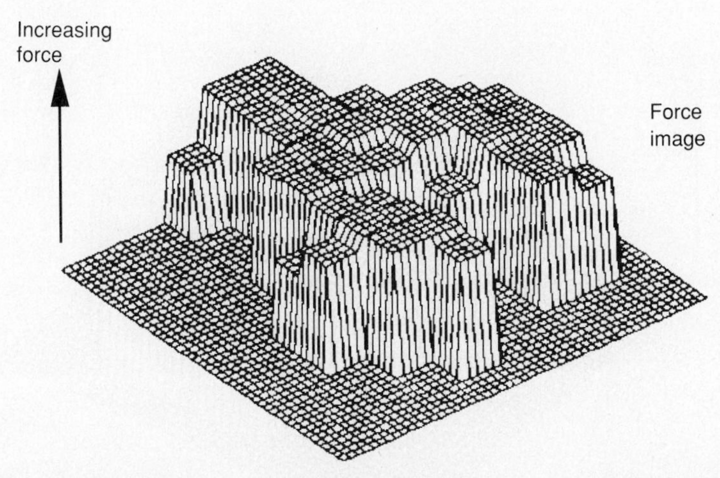


**Figure 5.35** An example of a force image produced by pressing a slotted electrical lug against a 10 x 10 array touch sensor

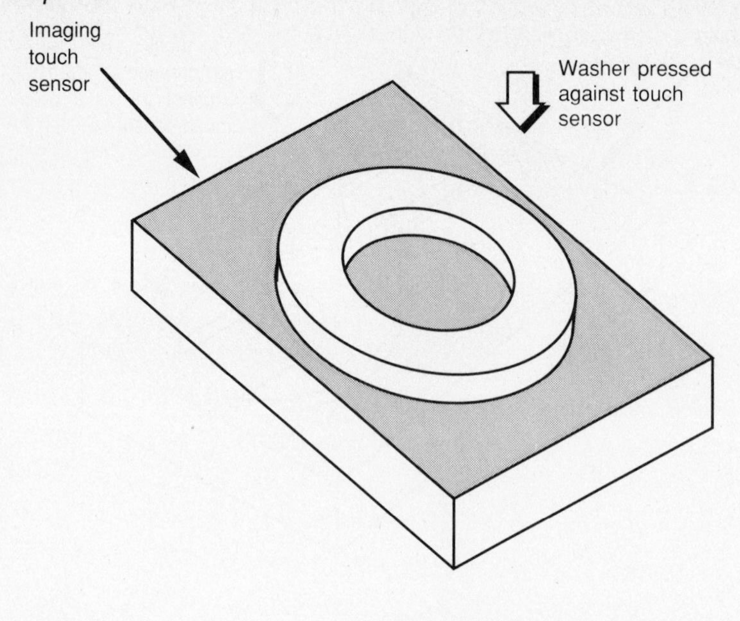


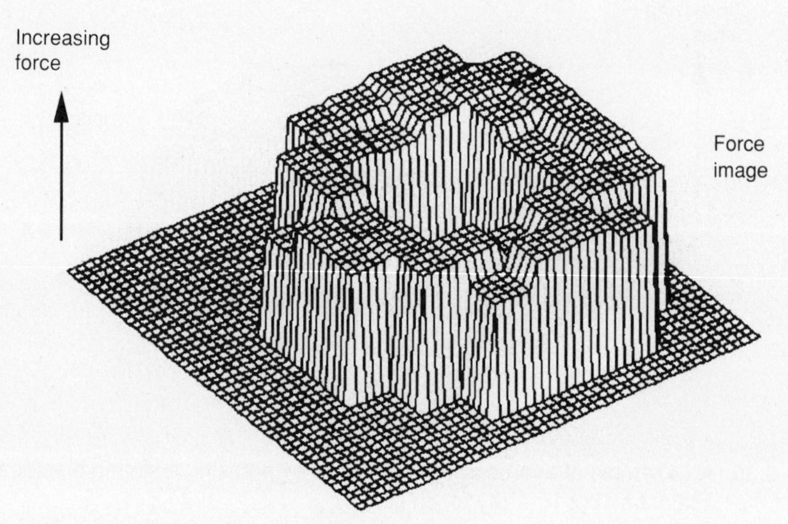


**Figure 5.36** An example of a force image produced by pressing a washer against a 10 x 10 array touch sensor

# Bibliography

Begej, S., 'Fingertip-shaped Optical Tactile Sensor for Robotic Applications', *Proceedings of the IEEE International Conference on Robotics and Automation*, Philadelphia, 24–29 April, 1988, pp. 1752–7.

Brady, M., 'Forward', *The International Journal of Robotics Research*, Vol. 7, No. 6, December 1988, pp. 2–4

CAMERON, A., et al. 'Touch and Motion', *Proceedings of the IEEE International Conference on Robotics and Automation*, Philadelphia, 24–29 April 1988, pp. 1062–7.

DARIO, P., et al, 'Tendon Activated Exploratory Finger With Polymeric Skin-Like Tactile Sensor', *Proceedings of the IEEE International Conference on Robotics*, 1985, pp. 701–6.

DARIO, P., and BUTTAZZO, G., 'An Anthropomorphic Robot Finger for Investigating Artificial Tactile Perception', *International Journal of Robotics Research*, Vol. 6, No. 3, Fall 1987, pp. 25–48.

DARIO, P., et al. 'Advanced Rehabilitative Robots', *Proceedings of the International Symposium and Exposition on Robots*', Sydney, Australia, 6–10 November 1988, pp. 687–703.

DE ROSSI, D., NANNINI, A., and DOMENICI, C., 'Artificial Sensing Skin Mimicking Mechanoelectrical Conversion Properties of Human Dermis', *IEEE Transactions on Biomedical Engineering*, Vol. 35, No. 2, February 1988, pp. 83–92.

DE ROSSI, D., et al., 'Tactile Sensing by an Electromechanical Skin Analog', *Sensors and Actuators*, Vol. 17, 1989, pp. 107–14.

EGHTEDARI, F., and MORGAN, C., 'A Novel Tactile Sensor for Robot Applications', *Robotica*, Vol. 7, 1989, pp. 289–95.

ERNST, H. A., 'MH–1 — A Computer-Operated Mechanical Hand', *Proceeding of the AFIPS Spring Joint Computer Conference*, Vol. 21, 1962, pp. 39–51.

GARRISON, R. L., and WANG, S. S. M., 'Pneumatic Touch Sensor', *IBM Technical Disclosure Bulletin*, Vol. 16, No. 6, November 1973, pp. 2037–40.

GRAHN A. R., and ASTLE, L., 'Robotic Ultrasonic Force Sensor Arrays', *Proceedings of the Robots 8 Conference*, Detroit, Michigan, 4–7 June 1984, pp. 21.1–21.18.

HACKWOOD, S., et al., 'A Torque-Sensitive Tactile Array for Robotics', *The International Journal of Robotics Research*, Vol. 2, No. 2, Summer 1983, pp. 46–50.

HARMON, L. D., 'Automated Tactile Sensing', *International Journal of Robotics Research*, Vol. 1, No. 2, Summer 1982, pp. 3–32.

HILLIS W. D., 'Active Touch Sensing', *MIT AI Memo* 629, April 1981.

KING, A. A., and WHITE, R. M., 'Tactile Sensing Array Based on Forming and Detecting an Optical Image', *Sensors and Actuators*, Vol. 8, 1985, pp. 49–63.

*Kynar Piezo Film Technical Manual*, Pennwalt Corporation, 1987.

LARCOMBE, M. H. E., 'Carbon Fibre Tactile Sensors', *Proceedings of the First International Conference on Robot Vision and Sensory Controls,* IFS (Publications) Ltd, Bedford, UK, 1981, pp. 273–6.

LIBOVE, C., 'Elastic Stability', in *Handbook of Engineering Mechanics*, W. Flügge, ed, McGraw-Hill Book Company, 1962, pp. 44.1–44.42.

LUO, R.-C., et al. 'An Imaging Tactile Sensor with Magnetostrictive Transduction', *Proceedings of the First International Conference on Intelligent Sensors and Computer Vision*, Cambridge, Massachusetts, 1984.

MAALEJ, N., and WEBSTER, J. G., 'A Miniature Electrooptical Force Transducer', *IEEE Transaction on Biomedical Engineering*, Vol. BME-35, No. 2, 1988, pp. 93–8.

MORRISH, A. H., *The Physical Principles of Magnetism*, John Wiley & Sons, Inc., New York, 1965.

MOTT, D. H., et al., 'An Experimental Very-High-Resolution Tactile Sensor Array', in *Robot Sensors*, Vol. 2, *Tactile and Non-Vision*, A. Pugh, ed, IFS (Publications) Ltd, Bedford, UK, 1986, pp. 179–88.

PETERSEN, K., et al. 'A Force Sensing Chip Designed for Robotic and Manufacturing Automation Applications', *IEEE International Conference on Solid-State Sensors and Actuators*, 1985, pp. 30–2.

PURBRICK, J. A., 'A Force Transducer Employing Conductive Silicone Rubber', *Proceedings of the First International Conference on Robot Vision and Sensory Controls*, IFS (Publications) Ltd, Bedford, UK, 1981, pp. 73–80.

RAIBERT, M. H., 'An All Digital VLSI Tactile Array Sensor', *Proceedings of the IEEE International Conference on Robotics*, 1984, pp. 314–9.

REBMAN, J., and MORRIS, K. A., 'A Tactile Sensor with Electro-optical Transduction', in *Robot Sensors*, Vol.2, *Tactile and Non-Vision*, A. Pugh, ed, IFS (Publications) Ltd, Bedford, UK, 1986, pp. 145–55.

SCHNEITER, J. L., and SHERIDAN, T. B., 'An Optical Tactile Sensor for Manipulators', *Robotics and Computer-integrated Manufacturing*, Vol. 1, No. 1, 1984, pp. 65–71.

SCHOENWALD, J. S., et al. 'A Novel Fiber Optic Tactile Array Sensor', *Proceedings of the IEEE International Conference on Robotics and Automation*, Raleigh, North Carolina, 31 March–3 April 1987, pp. 1792–7.

SIEGEL D. M., et al. 'A Capacitive Based Tactile Sensor', *SPIE Conference on Intelligent Robots and Computer Vision*, Cambridge, Massachusetts, September 1985, pp. 153–61.

SNYDER, W. E., and ST CLAIR, J., 'Conductive Elastomer as Sensor for Industrial Parts Handling Equipment', *IEEE Transactions on Instrumentation and Measurement*, Vol. IM-27, No. 1, March 1978, pp. 94–9.

TANIE, K., et al. 'A High-Resolution Tactile Sensor', in *Robot Sensors*, Vol.2, *Tactile and Non-Vision*, A. Pugh, ed., IFS (Publications) Ltd, Bedford, UK, 1986, pp. 189–98.

VRANISH, J. M., 'Magnetoresistive Skin for Robots', in *Robot Sensors*, Vol.2, *Tactile and Non-Vision*, A. Pugh, ed., IFS (Publications) Ltd, Bedford, UK, 1986, pp. 99–111.

WINGER, J., and LEE, K.-M., 'Experimental Investigation of a Tactile Sensor Based on Bending Losses in Fiber Optics', *Proceedings of the IEEE International Conference on Robotics and Automation*, Raleigh, North Carolina, 31 March-3 April 1988, pp. 754–9.

## Questions

5.1 Make a list of desirable capabilities and attributes for a touch sensor array.

5.2 Explain the differing characteristics you would expect from a tactile sensor which behaved like a pure force sensor and a pure pressure sensor.

5.3 Why is it important for a robot manipulator system to be able to detect shear forces at its fingertips?

5.4 Describe two touch sensor designs which are capable of measuring shear forces.

5.5 Compare the advantages and disadvantages of touch sensors based on piezoelectric polymer with the advantages and disadvantages of touch sensors based on conductive elastomer.

5.6 A piece of PVF2 polymer sheet 1 cm long by 1 cm wide by 28 μm thick, $\varepsilon_r = 12$, supports a mass of 1 kg resting evenly on its 1 $cm^2$ face. The wiring to a high impedance voltmeter has a capacitance of 100 pF. What change in voltage is registered by the meter as the mass is placed on the film?

5.7 Explain how a touch sensor based on frustrated internal reflection can be modified to measure shear forces.

5.8 Explain how a touch sensor can be built using microbending of an optical fiber.

5.9 Each of the sensor designs in this chapter employ an elastic material to convert applied load into a deflection. The deflection is then measured to provide the sensor output. Make a list of all the sensors described in this chapter and for each record the type of elastic material used (e.g., elastomer, silicon, steel, etc.).

5.10 The incompressibility of a solid elastomer skin presents a problem for those sensors which measure skin thickness. A large area of contact prevents radial expansion of the elastomer and thus much more force is required to indent an object into the sensor. The problem can be overcome by using elastomer containing voids which progressively collapse when a load is applied. List those sensing techniques which will not work with elastomer skins containing voids.

5.11 Which of the sensor techniques described in this chapter do you think is best suited to the construction of a practical robot tactile sensor for use in industry? Justify your choice by reference to the criteria suggested by Harmon (1982) and any other relevant considerations you can think of.

# Compliant tactile sensors

When a robot manipulator is positioned by dead reckoning alone there is good reason to grip objects between non-compliant surfaces. A compliant surface introduces some uncertainty about how an object is situated with respect to the fingers; however, there are advantages to be gained by having compliant gripping surfaces on a robot hand. A fingertip which molds around an object provides a more stable grip (Brockett 1985). This will be especially useful where the precise nature or position of the object is unknown and a trial lift is being performed. The cushioning effect of compliant gripping surfaces will also reduce the possibility of damage and wear to finger surface and gripped objects.

By adding sensing capabilities to a compliant fingertip the uncertainty about object position can be resolved. Thus, sensing can overcome a problem associated with a flexible skin. In addition, flexibility can improve the sensing performance of a tactile sensor.

## 6.1 How compliance affects the information-gathering capabilities of a tactile sensor

By definition tactile sensors must touch an object in order to derive any information. To some extent the larger the area of surface contact the greater is the amount of information which can be gathered. Very rigid tactile sensors produce detailed shape information for objects with extensive flat areas of contact. However, many domestic and industrial parts would only make a small number of point or line contacts with such a sensor. Figure 6.1 illustrates this point. A teaspoon only registers two isolated contact points.

It is informative to examine the relationship between sensor compliance, or rather the depth that an object can indent into the sensor (Figure 6.2), and the resulting area of surface contact.

Consider a range of probes consisting of a rod, radius $r$, shaped at one end to give either a cone, a hemisphere or a flat surface as shown in Figure 6.3. Assume that the skin material conforms perfectly to the probe surface. The relationships between indentation depth $x$ and area of surface contact (to a maximum depth of $r$) for these three probes are:

$$a_{cone} = \sqrt{2} . \pi . x^2 \qquad (6.1)$$

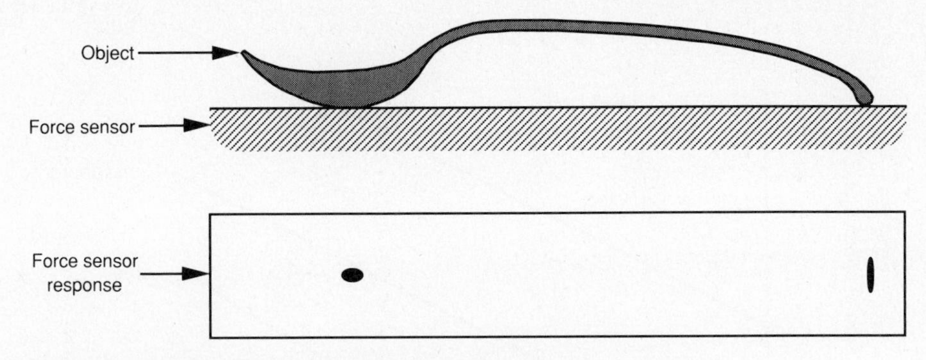


**Figure 6.1** Points of contact between a teaspoon and a rigid tactile sensor
(Redrawn from Russell 1988b, courtesy of Butterworth Scientific Ltd)

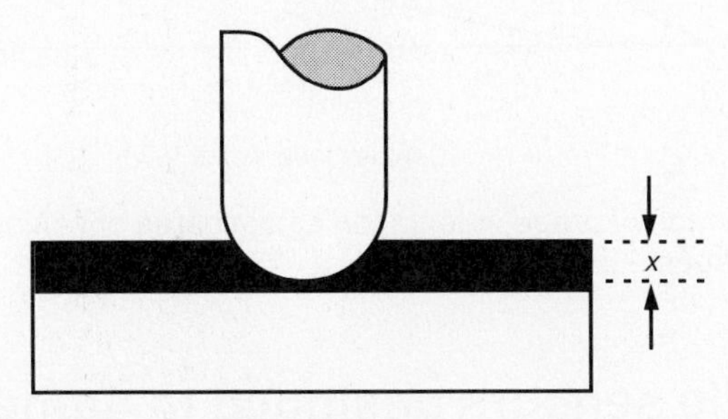


**Figure 6.2** Depth of indentation into a tactile sensor

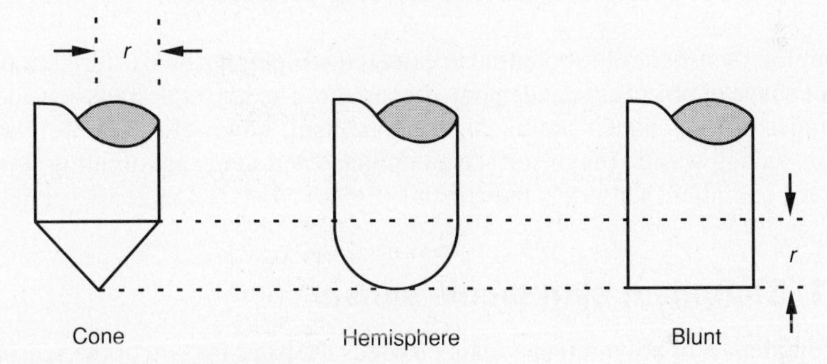


**Figure 6.3** Three probe tip shapes

$$a_{hemisphere} = 2.\,\pi.\,r.\,x \tag{6.2}$$

$$a_{blunt} = \pi.\,r^2 + 2.\,\pi.\,r.\,x \tag{6.3}$$

Figure 6.4 shows a plot of area of surface contact against depth of indentation for each of the probe tip shapes. This clearly shows that for all except very flat objects a rigid sensor provides little information. The deeper an object can indent into a tactile sensor, the more surface information can be obtained.

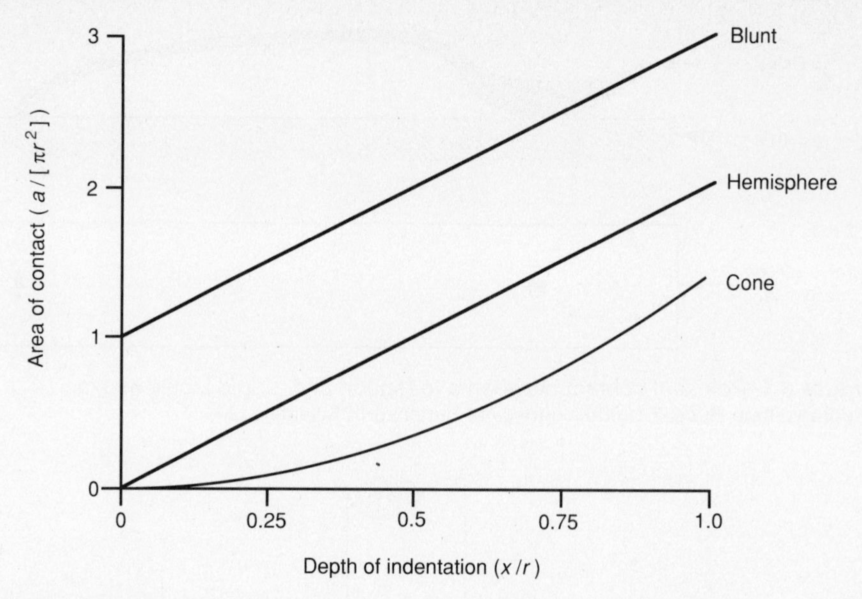


**Figure 6.4** A graph of probe indentation against area of contact for the three probe tip shapes

## 6.2 Tactile sensors designed to gather three-dimensional surface information

A compliant sensor has the potential to extract three-dimensional information about the surface shape of objects that indent into its surface. The sensor designs considered in the remainder of this chapter are all highly compliant; however, there are many sensor designs having a wide range surface compliance and the exact dividing line between rigid and compliant sensors is indistinct.

### 6.2.1 Compliant skin tactile sensor

The compliance of polyurethane foam provides the basis for two of the sensor designs. In this first design, the foam is covered with an outer skin of silicone rubber (see Figure 6.5). This structure molds round an object and yet returns to its original shape when the object is removed. Conductive rubber strain gauges are attached to the inner surface of the skin and deformation of the skin is registered as tension or compression in the gauges.

Because the polymer gauge material was only available in sheet form it was cut to the shape shown in Figure 6.6 and glued to the underside of the vulcanized silicone rubber skin. This configuration allowed individual gauge elements to be addressed using the potential divider technique and avoided making electrical connections in an area of the sensor subject to a high degree of bending and stretching.

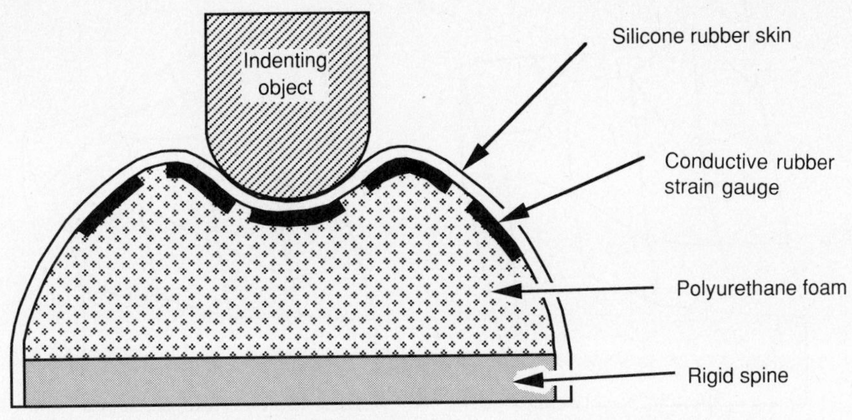


**Figure 6.5** The flexible skin tactile sensor
(Redrawn from Russell 1987, courtesy of MIT Press)

(a) Pattern of stretch-sensitive elastomer

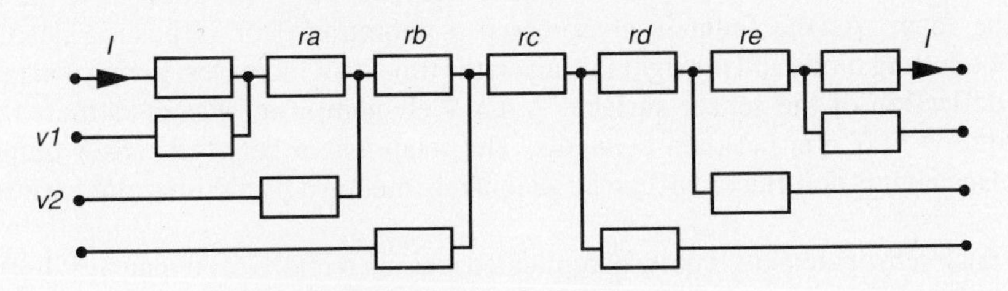


(b) Equivalent resistance of elastomer material

**Figure 6.6** Sensor elements made from stretch-sensitive elastomer

A 5 x 5 element array has been constructed with sensor spacing of 6 mm along the potential divider and 10 mm between adjacent potential dividers. Figure 6.7 shows results obtained with this sensor. From the plotted results it is possible to see differences in the radius of curvature of objects indented into the sensor. This sensor registers stresses in the sensor skin. A transformation must be performed on the sensor data to reconstruct the surface shape of an indenting object.

Because this sensor uses potential divider addressing, it contains a large number of

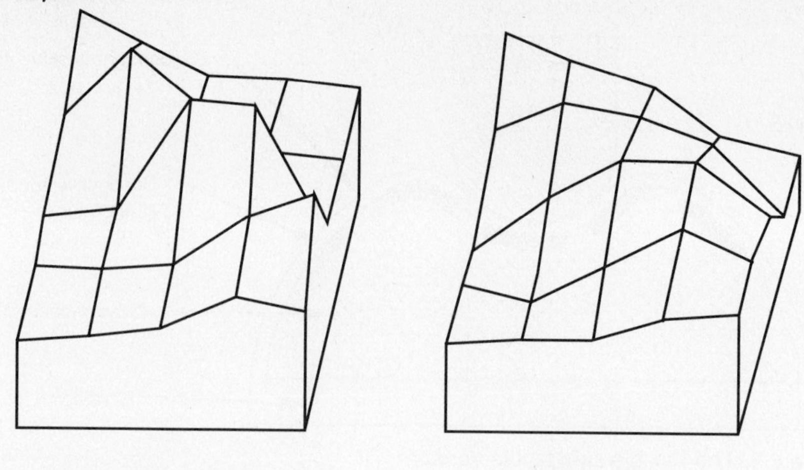


**Figure 6.7** Data from a 5 x 5 flexible skin sensor — strain increases vertically

interconnections (40 for the 5 x 5 sensor). If reliable connections could be made between the gauge material and elastomeric wires (flexible and having a low resistance) then row and column addressing could be implemented.

### 6.2.2 Compliant opto-mechanical pad sensor

This sensor also uses polyurethane foam to provide a highly compliant surface (see Figure 6.8). The thickness of the foam pad is measured by probes attached to the surface of the foam. As the foam is compressed a phototransistor is pushed nearer to a photoemitting diode and the light falling on the transistor increases, giving a measure of the deflection of the sensor surface. A 4 x 9 element array was constructed with a separation of 7 mm between elements. The relationship between sensor output and displacement is non-linear but can be adequately modeled by a third order least squares fit.

Each sensor element is quite complicated making it relatively expensive, bulky, and difficult to miniaturize. In its present form the mechanism would be damaged by large shear forces. However, the basic idea is promising and the design improvements could make this into a practical tactile sensor.

### 6.2.3 Parallel probe tactile transducers

These sensors consist of an array of probes which are lowered onto an object (see Figure 6.9). Deflection of the probes gives three-dimensional information about each point on the object which is touched by a probe.

Sato et al. (1986) describe an inductive technique for sensing probe deflection. A robot lowers an array of steel probes onto an object. As each rod touches and starts to ride up, an inductive sensor detects displacement of the probe. The movement of the

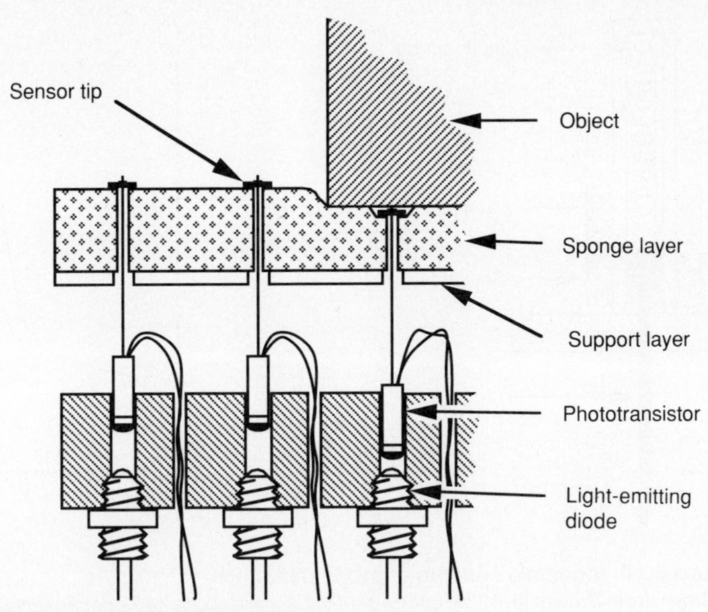


**Figure 6.8** A cross-sectional view of the flexible pad sensor
(Redrawn from Ozaki et al. 1982, ©1982 IEEE)

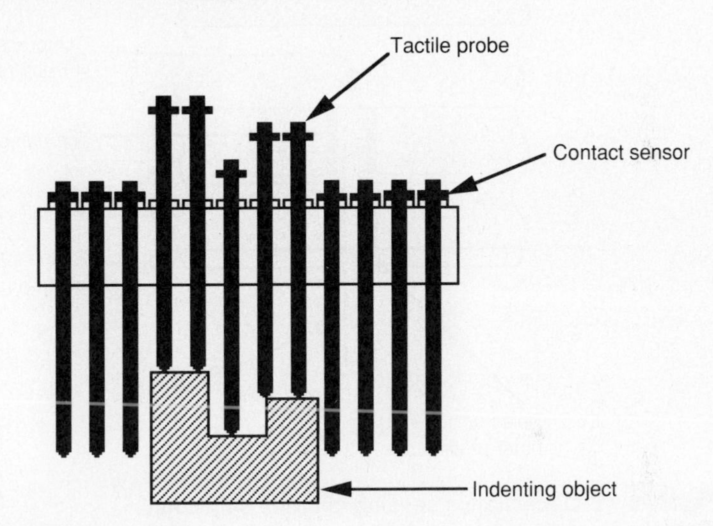


**Figure 6.9** The parallel probe sensor

robot coupled with the time at which a particular probe touched the object gives a three-dimensional point on the object. Probe movement is detected by measuring the electromagnetic coupling between two coils positioned at the end of each probe (Figure 6.10).

A capacitive technique for measuring probe deflection (Figure 6.11) is described by Jayawant (1986). The sensing probe forms part of a variable capacitor with a capacitance proportional to deflection. Capacitance is measured by exciting the capacitors with a linear voltage ramp and measuring the resulting current. Capacitor current:

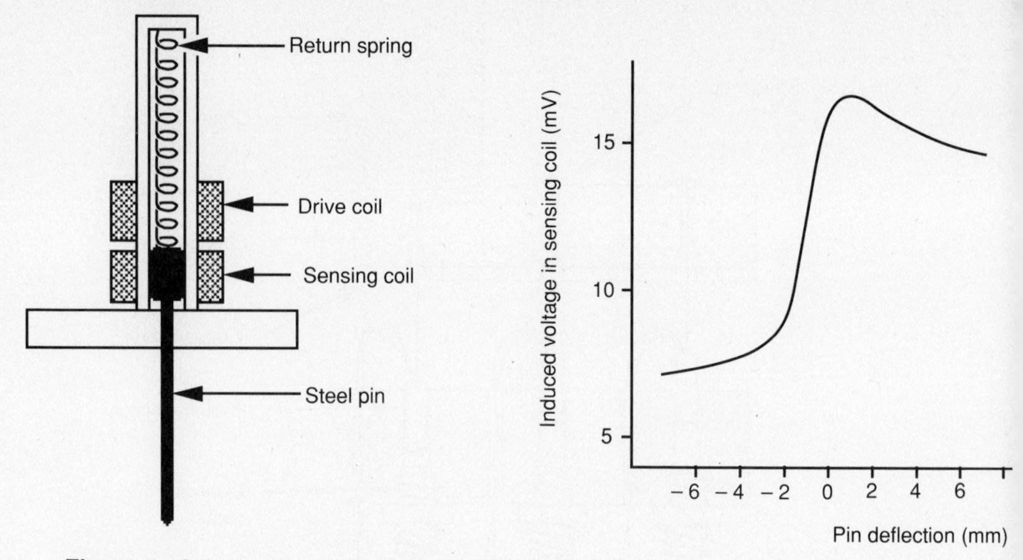


**Figure 6.10** Inductive sensing of probe deflection
(Redrawn from Sato et al. 1986, courtesy of the Japan Industrial Robot Association and the author)

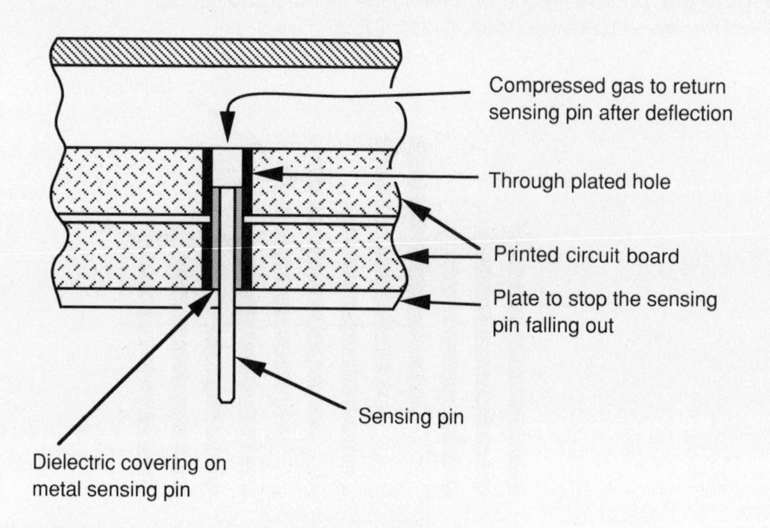


**Figure 6.11** Capacitative sensing of probe deflection
(Redrawn from Jayawant et al. 1986, courtesy IFS Publications)

$$i = C\frac{dv}{dt} \tag{6.4}$$

where:

$C$ = capacitance of the sensor element (farads);
$i$ = capacitor current (amperes); and

$\frac{dv}{dt}$ = rate of change of voltage across the capacitor which is constant for a linear voltage ramp.

Thus sensor capacitance is directly proportional to capacitor current. This technique has the advantage of providing an absolute measurement of probe deflection.

In general, parallel probes are prone to damage by bending, and stick if sideways forces are applied. But they can provide accurate three-dimensional surface contours of objects.

## 6.2.4 Three-dimensional probe

This mechanism (see Figure 6.12) is a three degrees of freedom positioning device, based on parallel actuation (Russell 1988a). Its function is to accurately position, by the control of winches, a tactile sensor in three-dimensional space. The winches, in turn, adjust the lengths of Kevlar tendons attached to the sensor. A conical compression spring maintains the tendons in tension and hence the sensor position depends only upon the position of three winches and the lengths of three tendons. This structure has the advantages of simplicity, accuracy, and resistance to damage. The compression spring ensures that the tendons are held in tension during all normal static and dynamic conditions. However, the system readily collapses under the effects of impacts or overloads, thus avoiding damage to the probe and preventing large forces being transmitted to external objects.

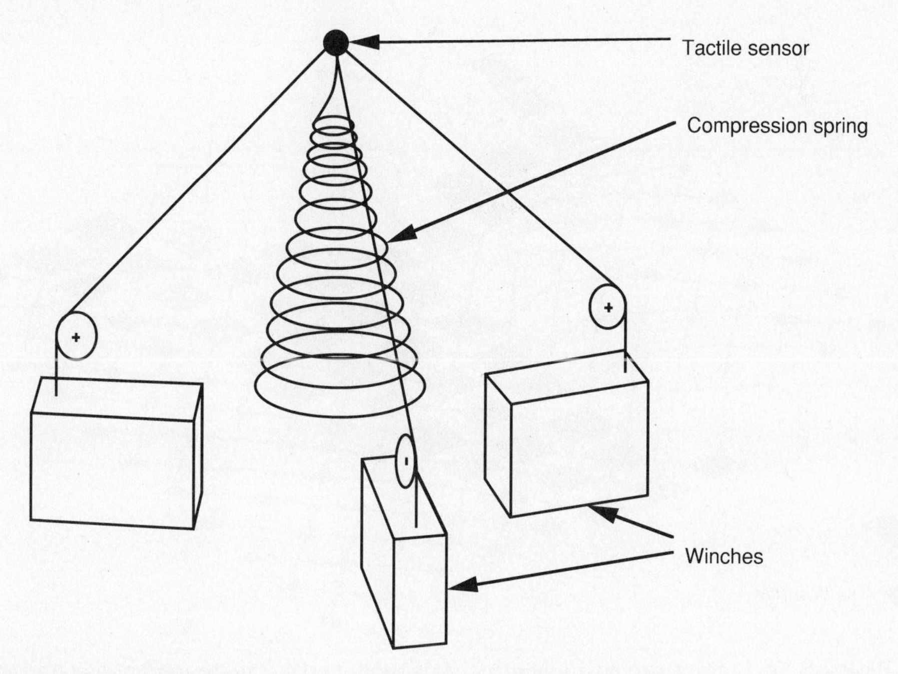


**Figure 6.12** A schematic diagram of the three-dimensional probe
(Redrawn from Russell 1988a)

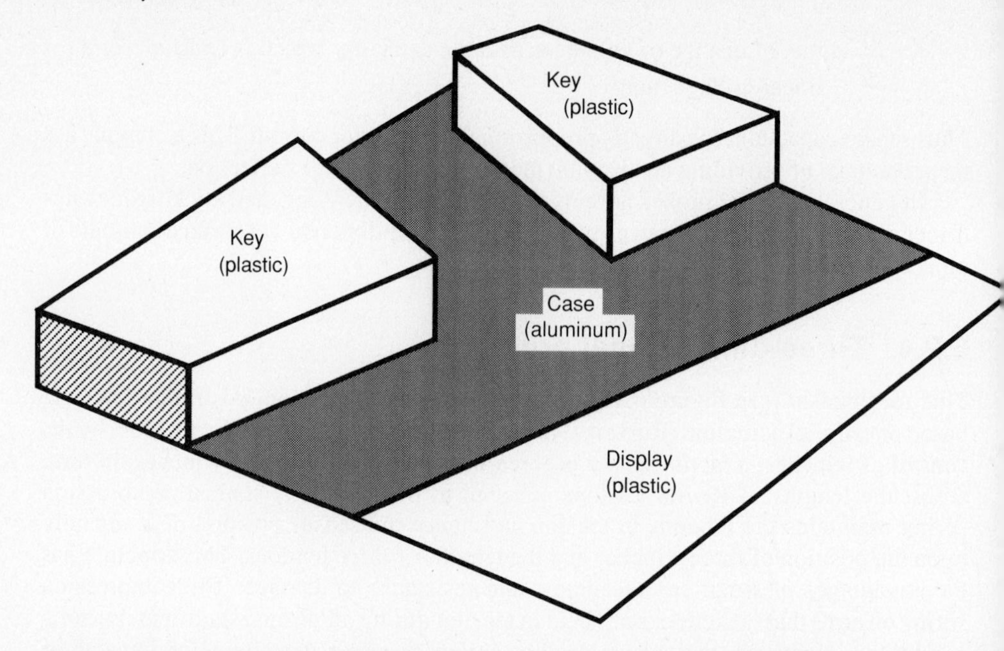


**Figure 6.13** A small area of the keyboard and display of a pocket calculator

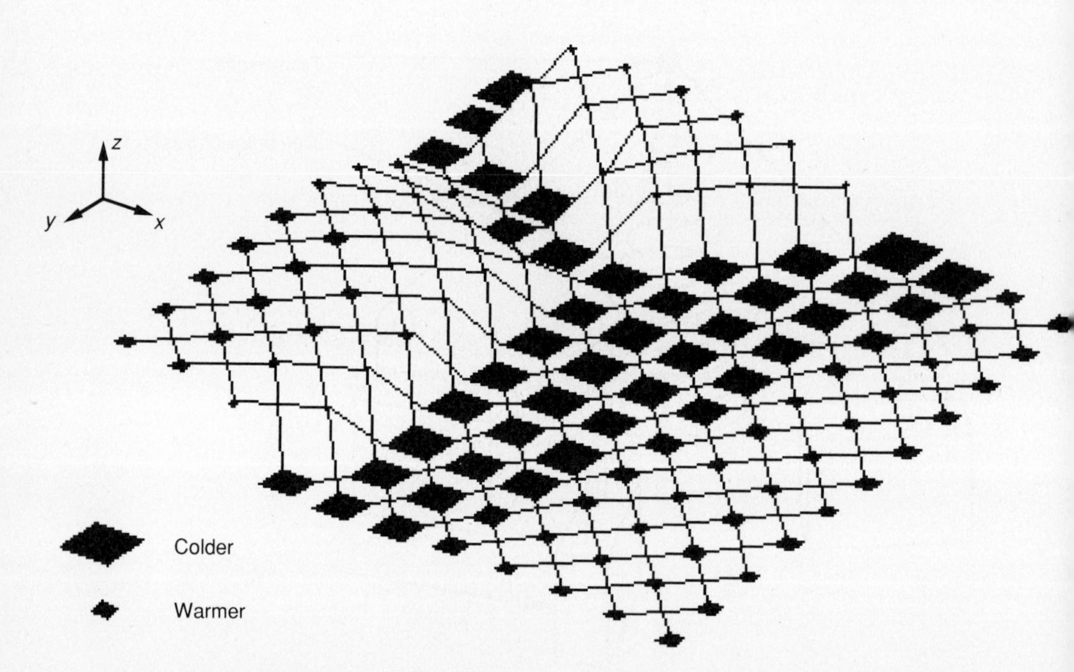


**Figure 6.14** Surface profile and thermal data produced by the three-dimensional probe scanning part of a calculator keyboard and display

It is envisaged that the tactile probe will carry sensors to measure one or more physical properties of a touched object. These properties may include temperature, thermal compliance, electrical conductivity, and surface contact.

Figure 6.13 shows a diagram representing a small area of the keyboard and display of a pocket calculator. This part of a calculator was scanned by a combined contact and thermal sensor mounted on the three-dimensional probe. The results given in Figure 6.14 show the surface profile and thermal information gathered by the sensor system.

## 6.2.5 Impedance tomographic tactile sensor

This sensor monitors the shape of a latex membrane by measuring the electrical conductivity of a volume of liquid contained beneath the membrane (see Figure 6.15). As an object indents into the membrane the conductive liquid (antifreeze containing ethylene glycol and other additives) is displaced. The prototype sensor contains a linear array of sixteen electrodes which are used to measure the resulting change in electrical conductivity. An alternating voltage source is applied across pairs of electrodes and the resulting voltages appearing between the remaining electrodes is measured. The volume of conductive liquid between two electrodes can be approximated by assuming a uniform cross-section $W$ wide by $H$ high and length $d$. Using this approximation, the resistance between electrodes is given by:

$$R = \frac{\rho \cdot d}{H \cdot W} \tag{6.5}$$

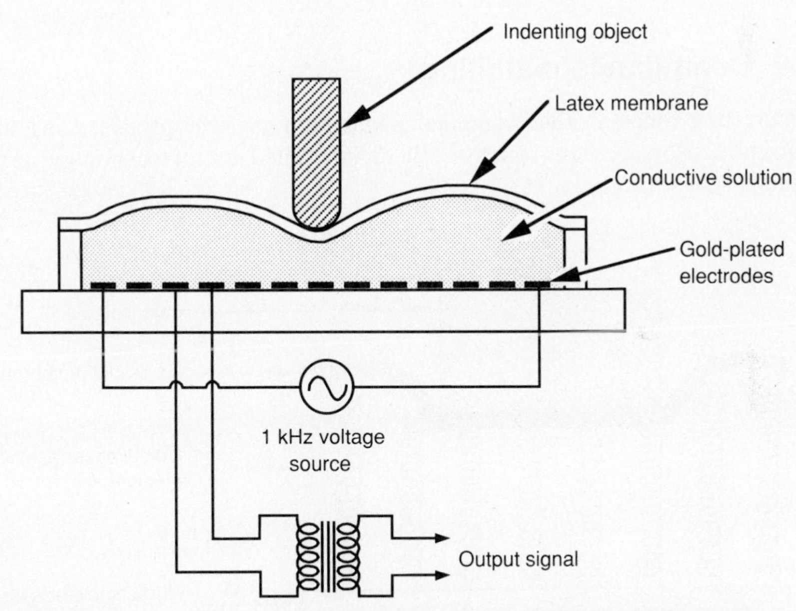


**Figure 6.15** Cross-sectional view of the impedance tomographic tactile sensor (Redrawn from Helsel et al. 1988, courtesy Elsevier Sequoia and author)

where:

$R$ = element resistance;
$\rho$ = resistivity of the antifreeze;
$d$ = separation between the electrodes;
$H$ = distance between electrodes and latex membrane; and
W = width of the sensor.

For the sensor reported by Helsel et al. (1988) $W = 1$ cm, $H = 2$ mm, and $d = 0.6$ mm. Assuming that a small indentation does not alter the total current flowing through the sensor, then:

$$\delta V = \frac{\delta H}{H} \cdot V \tag{6.6}$$

where:

$\delta V$ = change in voltage measured between two electrodes;
$\delta H$ = change in height of the latex membrane; and
$V$ = applied voltage.

Tests have shown a good correspondence between the predicted and observed electrode voltages. The prototype sensor only contained a line of sixteen electrodes. A two-dimensional array of electrodes will be required to determine the surface shape of an indenting object. By adding gelling agents to the conductive liquid, the compliance of the sensor may be modified to improve its gripping qualities.

### 6.2.6 Compliance matching

An interesting concept called compliance matching has been proposed as a means of using rigid sensors, perhaps based on silicon integrated circuit technology, as part of a compliant sensor (Figure 6.16).

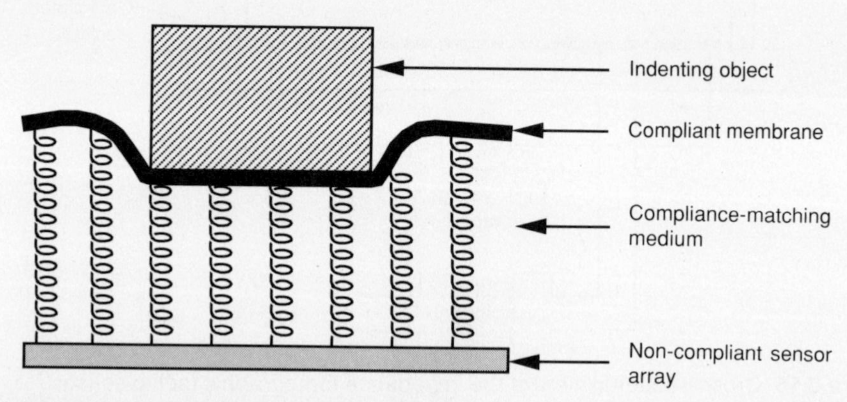


**Figure 6.16** The compliance-matching concept
(Redrawn from Clark 1988, ©1988 IEEE)

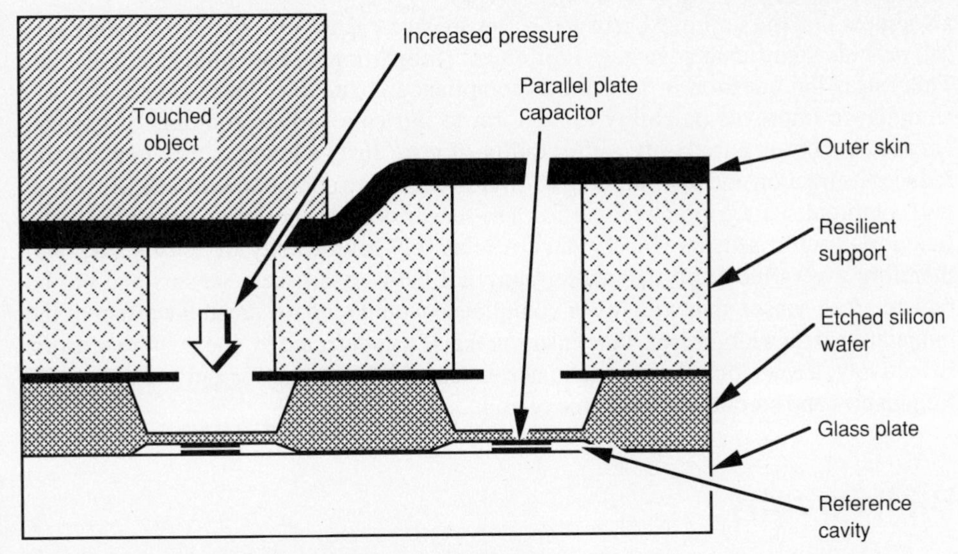


**Figure 6.17** The capacitive silicon tactile sensor
(Redrawn from Chun and Wise 1985, ©1985 IEEE)

A compliant membrane conforms to the shape of an indented object. The deformation of the membrane is transmitted to a non-compliant sensor array by means of a compliance-matching medium. Compliance matching may be achieved by means of contact using springs or foam rubber, or non-contact using electrostatic, magnetic, or light coupling.

Chun and Wise (1985) have made a sensor which uses air pressure to couple the deformation of a compliant outer skin to a relatively non-compliant capacitive sensor. This sensor can be viewed as an application of the compliance-matching concept. Figure 6.17 shows a cross-sectional view of the sensor. The sensor measures the deflection of a thin silicon membrane by detecting changes in the capacitance of a parallel plate capacitor which has one plate on the diaphragm and the other attached to the supporting glass plate. The diaphragm deflects due to a pressure difference between the reference cavity on one side and a compressible air chamber on the other. When an object indents into the sensor the air chamber is compressed and air pressure inside the chamber increases. This deflects the silicon diaphragm and increases the capacitance of the parallel plate capacitor. External circuitry measures the capacitance of each parallel plate capacitor and indirectly measures the profile of the sensor surface. Thus the resilience of compressed air provides compliance matching in this sensor.

## 6.3 Conclusion

In the past, little attention has been paid to the development of compliant sensors and hence there are fewer designs for this kind of sensor. It was felt that compliance introduced positional uncertainty and should be avoided. Yet there is an increasing awareness that the improved grip and better sensing capabilities of compliant sensors can provide significant advantages for object recognition and dextrous manipulation. This raises the question of how much compliance is suitable. Increasing a sensor's compliance improves its ability to conform to intricate surface contours of objects. However, there is a trade-off with stability of grip. Just as low pressure in a car tire reduces its traction and road-holding ability, so excessive compliance in a tactile sensor will compromise its gripping properties. The sensor skin also performs a useful filtering function. Tactile sensors usually have a relatively low density of sensor sites and therefore a very localized indentation may fall between adjacent sensors and not be registered. A sensor skin with lower compliance would tend to spread the effect of the indentation over a larger area, thus ensuring that it was detected by one or more sensors. Effectively, a low-compliance skin material acts as a filter, passing only lower spatial frequencies and smoothing the shape of an indented object.

## Bibliography

BROCKETT, R. W., 'Robot Hands with Rheological Surfaces', *IEEE International Conference on Robotics and Automation*, 1985, pp. 942–6.

CHUN, K. J., and WISE, K. D., 'A Capacitive Silicon Tactile Imaging Array', *IEEE International Conference on Solid-State Sensors and Actuators*, 1985, pp. 22–5.

CLARK, J. J., 'A Magnetic Field Based Compliance Matching Sensor for High Resolution, High Compliance Tactile Sensing', *Proceedings of the IEEE International Conference on Robotics and Automation*, Philadelphia, 24–29 April 1988, pp. 772–7.

HELSEL, M., et al., 'An Impedance Tomographic Tactile Sensor', *Sensors and Actuators*, Vol. 14, 1988, pp. 93–8.

JAYAWANT, B. V., et al., 'Robot Tactile Sensing: A New Array Sensor', in *Robot Sensors*, Vol. 2, *Tactile and Non-Vision*, A. Pugh, ed., IFS (Publications) Ltd, Bedford, UK, 1986, pp. 199–205.

OZAKI, H., et al., 'Pattern Recognition of a Grasped Object by Unit-Vector Distribution', *IEEE Transactions on Systems, Man and Cybernetics*, Vol. SMC-12, No. 3, May/June 1982, pp. 315–24.

RUSSELL, R. A., 'Compliant Skin Tactile Sensor', *IEEE International Conference on Robotics and Automation*, Raleigh, North Carolina, 1987, pp. 1645–8.

RUSSELL, R. A., 'A 3 Degree of Freedom Tactile Probe', *Proceedings of the International Symposium and Exposition on Robots*, Sydney, November 1988a, pp. 898–907.

RUSSELL, R. A., 'Optical Sensory Work Surface for a Robot Manipulator System', *Microprocessors and Microsystems*, Vol. 12, No. 9, November 1988b, pp. 527–31.

SATO, N., et al., 'A Method for Three-Dimensional Part Identification by Tactile Transducer', in *Robot Sensors*, Vol. 2, *Tactile and Non-Vision*, A. Pugh, ed., IFS (Publications) Ltd, Bedford, UK, 1986, pp. 133–43.

## Questions

6.1 Explain the advantages and disadvantages of a compliant tactile sensor compared to a rigid tactile sensor when used to provide sensory feedback for gripping and manipulation tasks.

6.2 Frictional force between two surfaces is proportional to the normal force and the coefficient of friction. How then does compliance reduce the tendency for slip between the two surfaces?

6.3 Obviously there must be some limit to the amount of compliance in a gripping surface. How much compliance do you think is reasonable? Justify your answer.

6.4 For each of the sensor designs described in this chapter explain how the surface profile of the indenting object could be calculated from the sensor output.

6.5 Explain the concept of compliance matching applied to tactile sensors.

# 7 Additional modes of tactile sensing

In Chapters 5 and 6 I have described tactile sensors which can determine contact forces and the surface profile of objects. Most of the interest in tactile sensing focuses on this group of sensors. However, other tactile sensing modes are available and in this chapter there are examples of sensors which employ these sensing modes. Some of the sensors measure a completely different physical property (such as thermal quantities); others directly measure a property which could be determined indirectly from the output of an array force/surface profile sensor by processing the sensor output (e.g., to determine slip).

## 7.1 Whiskers as proximity sensors

There are many applications for proximity sensing in robotics:

- Detecting the presence or absence of objects.
- Searching for small parts to be assembled. If a robot does not have sensory feedback, components must be offered to the robot in a precise orientation. The part feeders used for presenting components in a repeatable manner are an expensive and inflexible part of a robot installation.
- Protecting against damaging collisions. Even in a highly structured environment where object location is presumably known, accidental misplacement or presence of foreign objects or people would pose a hazard. This problem will increase as robots are used in less structured factory environments.

Although many techniques are available for detecting objects at close range, they all have limitations. Most proximity sensors based on reflected light beams and ultrasonic sonar are dependent on the reflectivity, surface roughness, and material of the sensed object. In common with fluidic sensors they are also sensitive to the relative orientation between the sensor and sensed object. Fluidic proximity sensors rely upon a flow of fluid to detect nearby objects. For example, as an object approaches a jet of air issuing from a nozzle the flow of air will be restricted. The resulting build-up of pressure in the nozzle can be measured and used to infer the presence of the object. Capacitive and inductive sensors are limited in range and their response depends on the material being sensed. Computer vision has great potential in the field of proximity sensing but systems available at present are expensive, require a lot of computing power to analyze the images, and are sensitive to surface finish of components, ambient lighting, and cluttered environments. Laser time-of-flight range-finding has also been considered for robot sensing but the equipment required is bulky and expensive. All of these proximity-sensing techniques would offer acceptable performance in certain situations but none fill the role of a general purpose multi-use proximity sensor.

Because whisker sensors are light and have low inertia they will not displace any but the lightest objects that they touch. Whisker sensors detect the presence of objects before they touch the surface of the organism/robot and for this reason they can be thought of as proximity sensors. A brief look at whisker and antennae sensors in nature will demonstrate some of their potential applications.

### 7.1.1 Biological whiskers

Nocturnal animals, insects, marine creatures, and even some plants place great reliance on whiskers or antennae to warn of danger, detect their prey or otherwise sense their environment. The domestic cat is one of the few animals where the sensing capabilities of whiskers has been studied in any detail. A cat captures and manipulates its prey using its mouth. Objects held in the mouth are too close for the cat's eyes to focus on and are also partially obscured from view. Whiskers allow the cat to determine the position and movements of prey animals close under its nose or held between its jaws (Figure 7.1). The upper lip of the cat is well supplied with whiskers (vibrissae). Each whisker is provided with about 200 sensory nerve endings which respond to displacement and velocity stimuli. Muscles position the whiskers to sense objects close to the mouth of the cat or fold them back to keep them out of the way while eating.

When springing onto its prey, or when carrying something, the whiskers are angled far forward to envelop the object. A blindfolded cat can locate a mouse. As soon as the

**Figure 7.1** A cat's whiskers envelop its prey

mouse touches the cat's whiskers the cat grasps the mouse with a fast and precise bite to the nape of the neck. There is also evidence that the cat uses its whiskers to find the direction of hairs on a prey animal's body which is an aid to cutting up and eating the animal.

Many insects have a hard keratinous external skeleton which means that any tactile sensors positioned beneath this tough layer will be relatively insensitive. The solution that insects have adopted is to mount their environmental sensors on whiskers projecting from their body. Many of these whiskers are sensitive to touch and detect contact with external objects, or are positioned at limb joints and determine the posture of the insect. Other whiskers, particularly those situated on the antennae are sensitive to:

- *air flow* — helping to control flight;
- *chemicals* — detecting pheromones for mating; chemical signals for communication in social insects such as ants, bees and termites; the scent of prey for fleas, mosquitoes and ticks; and tasting food before it is eaten; and
- *temperature* — sensed for protection and comfort.

The antennae, with their array of sensors, establish a clear volume for the insect to navigate, check its footing by tapping the ground ahead, and determine the presence of food or dangerous chemicals.

### 7.1.2 Technological whiskers

Like insects, robots have hard skins and can also benefit from the sensory information which whiskers provide.

#### *7.1.2.1 Simple whisker sensor*

A short length of piano wire passing through a small hole in a copper plate forms the basis of an inexpensive whisker sensor (Figure 7.2). When the wire is deflected it touches the copper plate thus completing an electrical circuit to signal the contact. A sensor like this has been used to locate and pick up randomly placed objects using a robot arm (Russell 1984).

#### *7.1.2.2 Pneumatic whisker sensor*

A similar whisker sensor has been described by Wang and Will (1978). Their whisker sensor (see Figure 7.3) has the additional refinement of being mounted on a pneumatic actuator so that the whisker can be retracted for protection when heavy lifting is performed.

#### *7.1.2.3 Whisker sensors for mobile robots*

Whisker sensors have many applications in the area of mobile robotics. Tufts of simple whisker sensors can be mounted over areas of a robot surface which are to be protected from collision. The walking robot Titan III employs whisker sensors positioned around the soles of its feet (Figure 7.4). The whiskers are made of shape-memory alloy because its high elasticity allows this material to tolerate a relatively large amount of bending

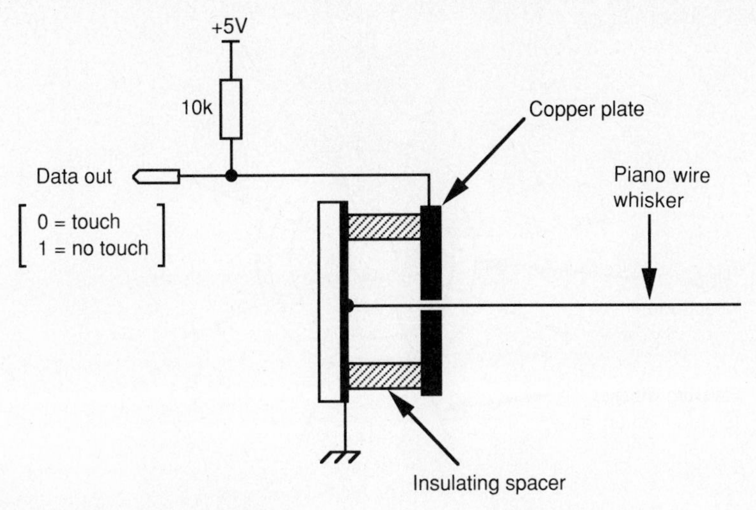


**Figure 7.2** A simple whisker sensor
(Redrawn from Russell 1984, courtesy of North American Technology Inc.)

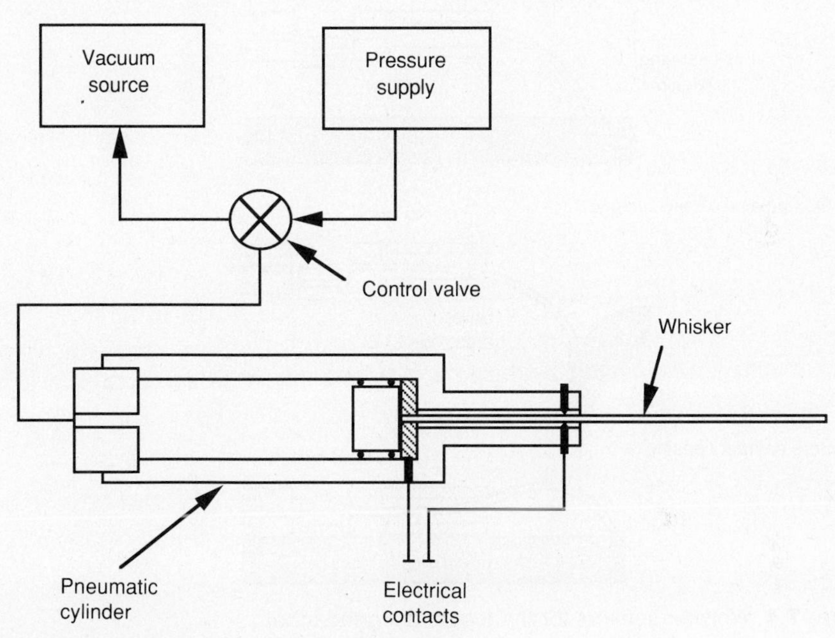


**Figure 7.3** Pneumatically retracted whisker probe
(Redrawn from Wang and Will 1978, courtesy IFS Publications)

without suffering permanent deformation. As the foot is lowered the whiskers detect the ground so that the foot can be decelerated before contact. During forward movement of the foot the whiskers also sense obstacles so that the foot can be lifted over them.

(1) Partial contact sensing

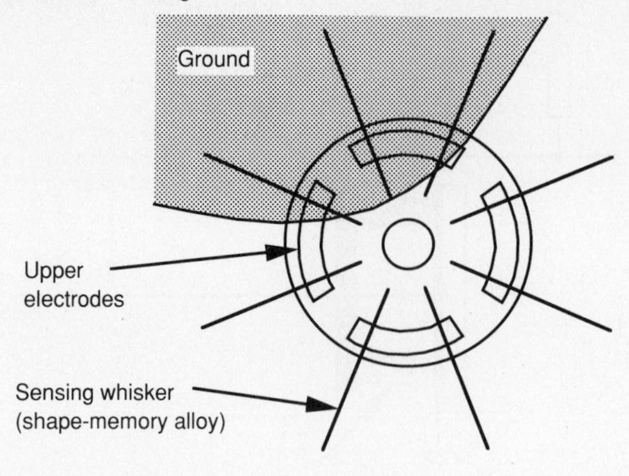


(2) Ground proximity sensing

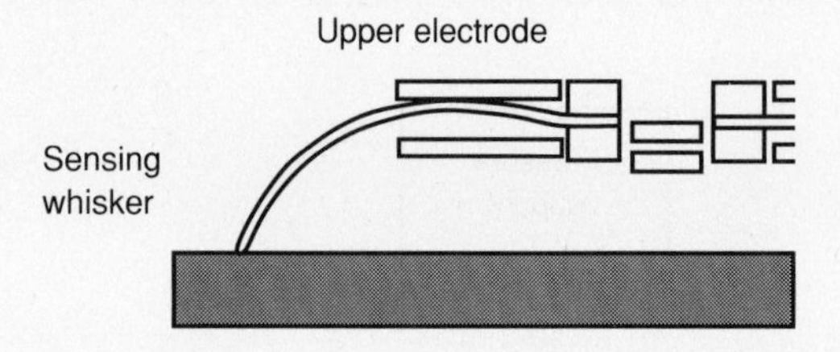


(3) Obstacle proximity sensing

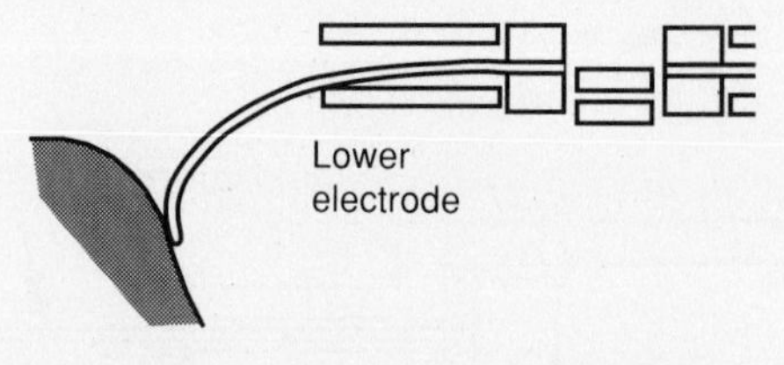


(4) Sole contact sensing

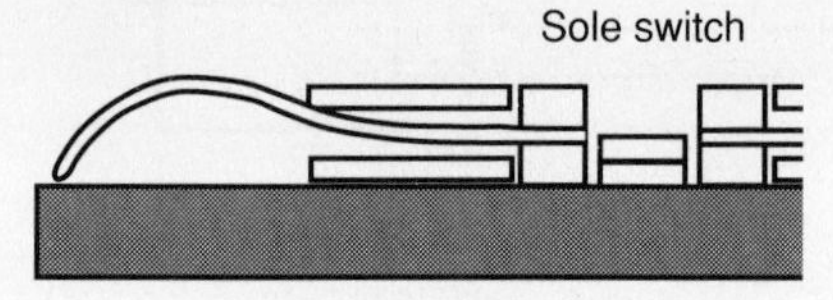


**Figure 7.4** Whisker sensors for the foot of a legged robot
(Redrawn from Hirose et al. 1985, ©1985 MIT Press)

### *7.1.2.4 Articulated whisker probe*

In order to gain sufficient information about an object outline to achieve recognition and to determine its orientation, several hundred measurements may be necessary. The inertia and response time of a robot manipulator makes this task a time-consuming

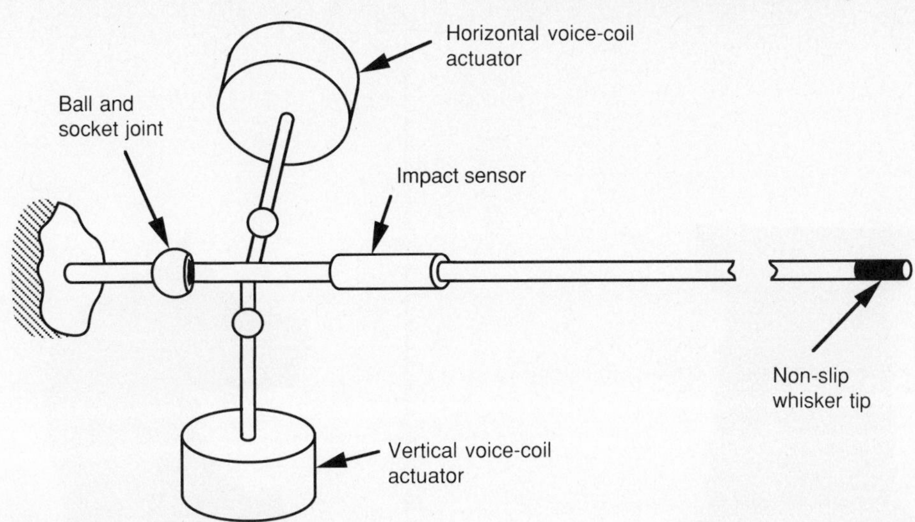


**Figure 7.5** An articulated whisker probe
(Redrawn from Russell, 1985)

process. In addition, any stiction or backlash in the robot drive chain will seriously degrade measurement accuracy. These problems can be reduced if the whisker is provided with its own servos to position the whisker in a similar manner to an insect's antennae. Figure 7.5 shows an articulated whisker sensor which uses this principle. Voice-coil actuators move the whisker in horizontal and vertical directions. An impact sensor (switch or piezoelectric crystal) detects contact between whisker and external object. This sensor can be used to build up silhouette images of objects by tracing their outline.

Figure 7.6 shows outline images of two chess pieces produced by an articulated whisker probe.

Whiskers are so inexpensive that large numbers of them can be mounted over the surface of the robot manipulator to warn of impending collisions. In this role they are effective during low-speed manipulation operations but have insufficient range to provide comprehensive protection during high-speed positioning maneuvers.

#### *7.1.2.5 Tactile seam tracking for arc welding*

Many industrial robots are employed in arc welding applications. To weld two metal pieces together, the welding torch must track the seam between them. In general, this tracking cannot be performed by dead reckoning because of variations in the precise shape of the work pieces. The environment close to a welding torch is very hostile with a broad spectrum of intense light extending from infra-red to ultra-violet wavelengths, electromagnetic interference, and flux fumes. Tactile sensing has been used to track the welding seam close to the welding torch, and an example of a tactile seam tracker is shown in Figure 7.7. This device contains an absolute optical encoder to measure side-to-side displacement of a pivoted tactile probe. The probe designers claim that it is fast, rugged, and accurate to within $\pm 0.5$ mm.

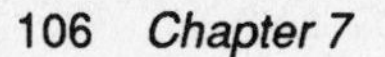

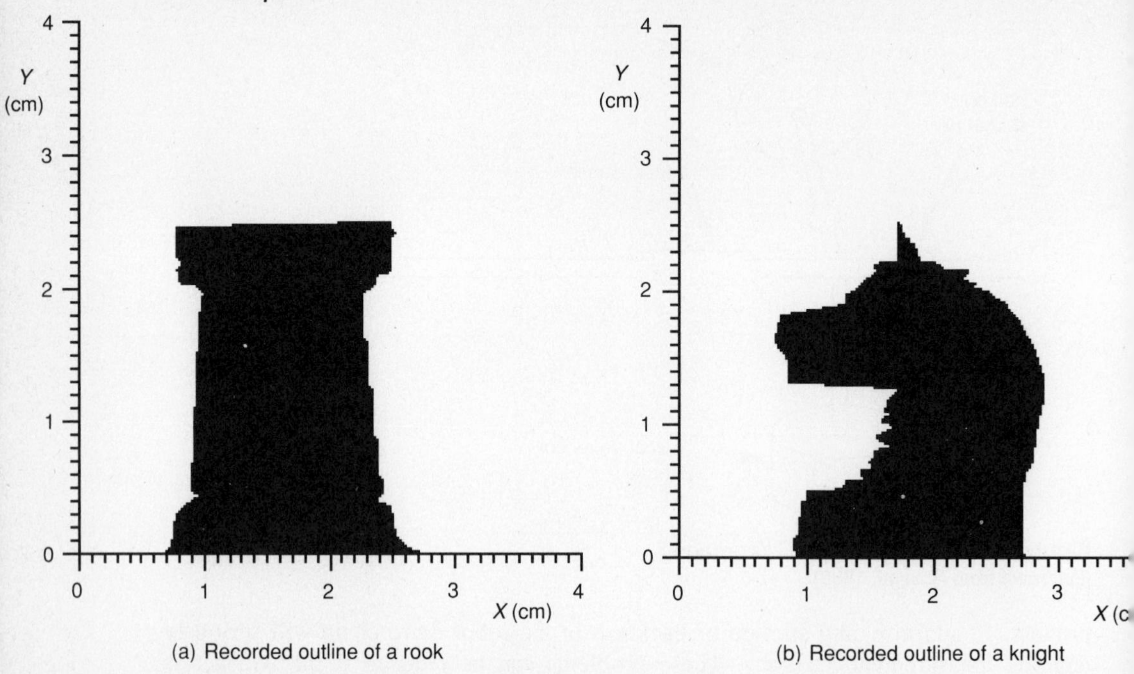


**Figure 7.6** Outline images produced by an articulated whisker probe
(Redrawn from Russell 1985)

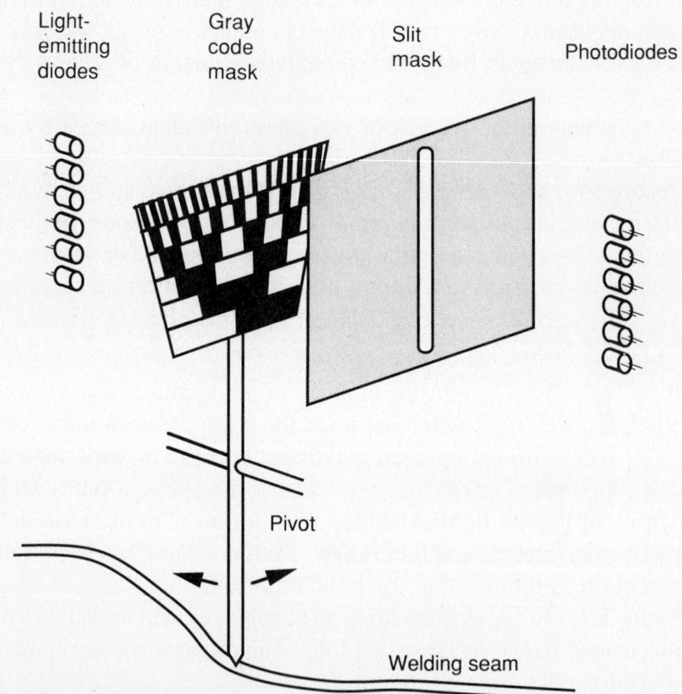


**Figure 7.7** Tactile weld seam tracker
Redrawn from Presern et al. 1981, courtesy IFS Publications)

## 7.2 Thermal sensing (Russell 1984)

Robots are less sensitive to extremes of temperature compared to their human counterparts, and perhaps this may explain why little attention has been paid to the development of thermal sensors for robots. As more sensors and their associated electronics are built into robot grippers it will become necessary to provide protection against damaging temperatures. In addition, humans use their temperature sense to determine the thermal properties of the objects they touch. Information about the thermal properties of objects will also be useful for autonomous robot systems. For these reasons a robot thermal sense would be useful for a robot to survive and recognize unknown objects in an unstructured environment.

A thermal sensor can be modeled on the human thermal touch sense. The skin houses capillary blood vessels which are one of the main sources supplying heat to the skin and regulating that supply. The fact that the skin is usually maintained at a temperature well above ambient provides a mechanism for assessing the thermal properties of touched objects. From our own experience we know that steel feels colder than wood even though both materials are at room temperature. The apparent difference is caused by steel conducting heat from our fingers quicker than wood. Under these circumstances the sensations of cold and warmth provide information about the material constitution of the touched object.

Therefore, there are three main parts to a thermal sensor:

1. A temperature stabilized heat source to warm the sensor in the same manner that the blood supply warms the skin.
2. A layer of material of known thermal conductivity to couple the heat source to the touched object.
3. An array of temperature transducers to measure the contact point temperature over the sensor surface. These transducers replace the thermally sensitive nerve endings in the skin.

Figure 7.8 shows a schematic diagram of such a thermal sensor. Heat dissipated in a power transistor warms the sensor. The heater temperature is stabilized by feedback from thermistor TH2. Thermistor TH1 measures the temperature drop caused by contact with an external object and the voltage across this thermistor forms the sensor output. The temperature sensed by TH1 varies with time after contact with an external object. Materials of differing thermal conductivity and diffusivity produce distinct and reproducible time-varying responses from the sensor as illustrated in Figure 7.9.

This graph shows the evolution of sensor temperature when samples of four different materials are applied to the sensor five seconds after the start of recording and removed twenty seconds later. These results agree with our own experience. Aluminum feels cold, which corresponds to the large temperature drop on the graph. Paraffin wax, cork and polystyrene foam feel progressively less cold and give correspondingly smaller temperature drops. In fact, polystyrene foam is such a good insulator that the sensor temperature rises above its value in air after a short period of contact with the foam.

How can the sensor output be analyzed to determine the material in contact with the sensor? The sensor gives a markedly different response for each material tested. This

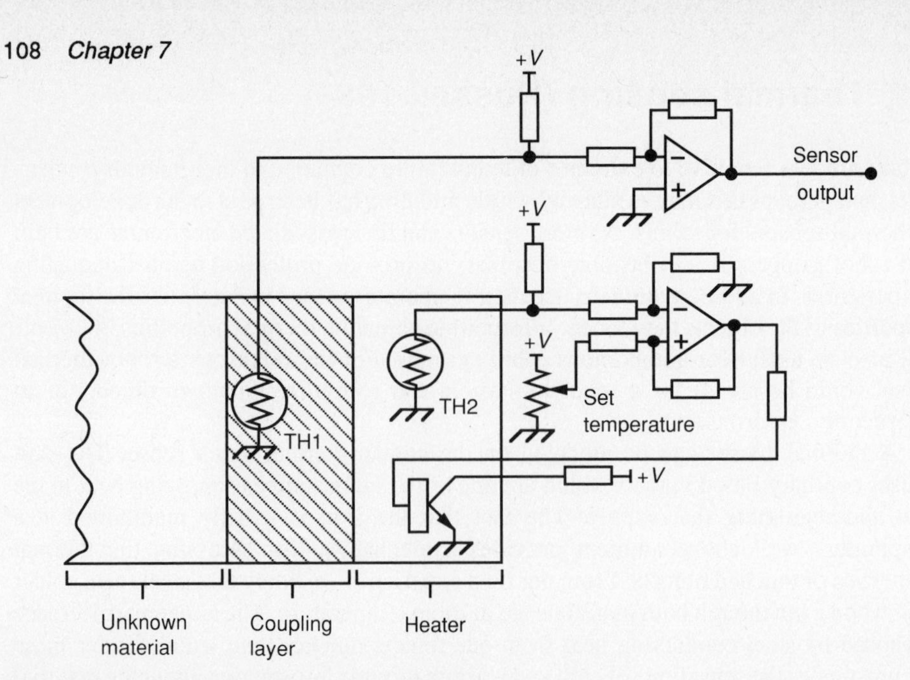


**Figure 7.8** A schematic diagram of a thermal sensor
(Redrawn from Russell 1985, ©1985 MIT Press)

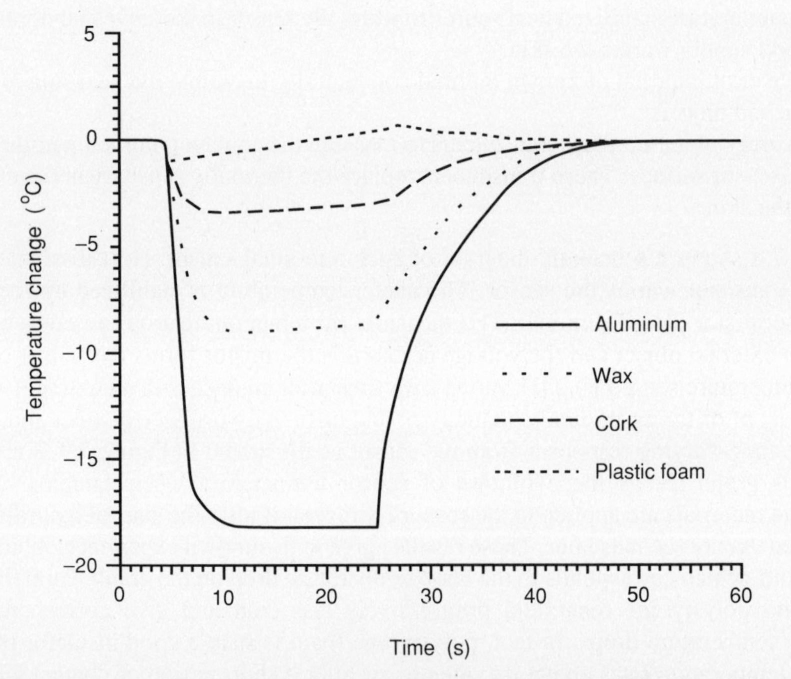


**Figure 7.9** Thermal sensor output: sensor temperature change for four materials showing the effect of contact with the materials followed by separation
(Redrawn from Russell 1985, ©1985 MIT Press)

difference is apparent after one or two seconds and therefore a reading from the sensor shortly after contact gives information which will allow different materials to be distinguished. In order to provide a measurement which is independent of the ambient and initial temperatures the percentage sensor temperature drop towards ambient is calculated a fixed time after contact:

$$D = \frac{T_i - T_c}{T_i - T_a} .100 \qquad (7.1)$$

where:

$D$ = percentage temperature drop towards ambient;
$T_i$ = initial sensor temperature;
$T_a$ = ambient temperature; and
$T_c$ = sensor temperature after contact.

The percentage temperature drop towards ambient can be used to categorize an unknown material in terms of a range of prerecorded materials. Values of $D$ are recorded for a number of materials having widely varying thermal properties. Figure 7.10 shows the values of $D$ recorded for samples of aluminum, glass, wax, perspex, wood, and cork. Unknown materials are then categorized in terms of the known material with the closest $D$ value.

By extending the thermal sensor into a two-dimensional array, images can be produced which show the shape and different materials making up an object (Figures

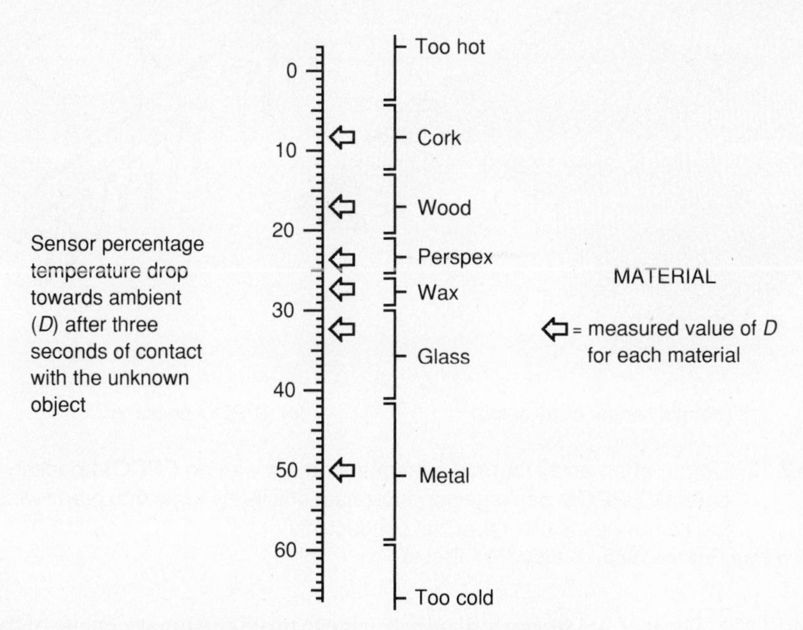


**Figure 7.10** Prerecorded values of *D* for a range of materials
(Redrawn from Russell 1984, courtesy North American Technology Inc.)

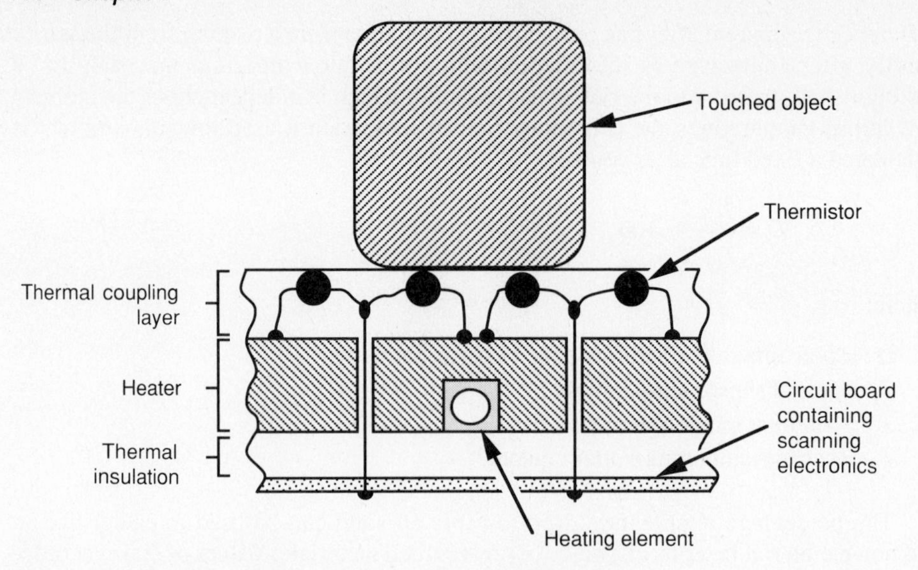


**Figure 7.11** Cross-sectional view of an array thermal sensor
(Redrawn from Russell 1988, courtesy Cambridge University Press)

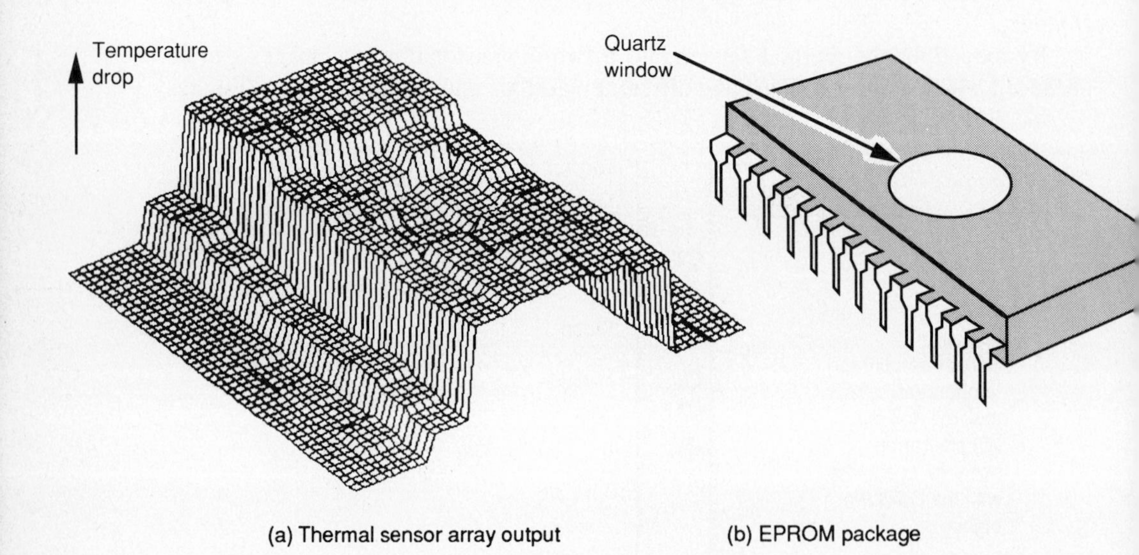


(a) Thermal sensor array output (b) EPROM package

**Figure 7.12** Output of an array thermal sensor in contact with an EPROM package. The ceramic EPROM package conducts heat efficiently while the quartz window in the center has a lower thermal conductivity
(Redrawn from Russell 1985, courtesy MIT Press)

7.11 and 7.12). The thermal sensor will only be able to provide a binary image of the area of contact with an external object and, because of thermal conduction across the face of the sensor, the edges of the image will not be sharply defined.

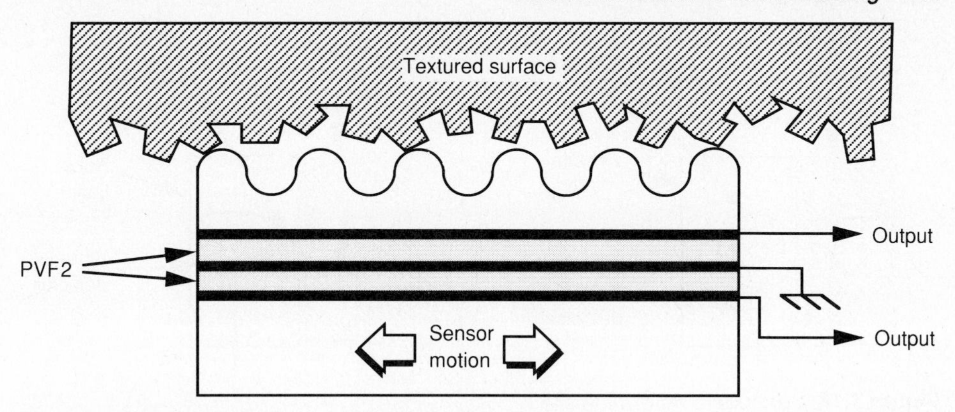


**Figure 7.13** Cross-sectional view of a texture sensor
(Redrawn from Patterson and Nevill 1986, courtesy Cambridge University Press and authors)

## 7.3 Texture sensing

The exact point at which surface irregularities become so small that they constitute a texture is open to argument. I think of texture sensors as devices which 'play' the surface of an object rather like a gramophone needle playing a record. The human texture sense works in this manner — the finger must be drawn across a surface to determine its texture. Patterson and Nevill (1986) have designed a texture sensor incorporating a ridged surface pattern (Figure 7.13). This pattern creates vibrations when the sensor is scanned across a surface. Vibrations caused by the interaction of the ridged pattern with a surface texture are detected by piezoelectric polyvinylidene fluoride (PVF2) polymer sheets. The PVF2 sheet has a maximum sensitivity in one direction; therefore, to equalize the response, two crossed sheets are used. This sensor has no provision for mobility of its own and so uses a turntable to scan objects across the sensor at constant speed. The sensor output is analyzed for spectral content and has been used to discriminate different grades of sandpaper and spheres of varying diameters.

## 7.4 Slip sensing

The movement of an object held by a robot gripper, as it slips through the fingers, can be detected by an imaging force sensor. Analysis of the varying sensor output can show the direction and rate of slip. An alternative, more straightforward, method is to build a sensor whose output is directly related to slip. Ueda et al. (1972) designed a range of sensors for this purpose (Figure 7.14). The sensors work on two main principles:

1. As a gripped object slides through the fingers, a rubber roller pressing against the object rotates. The rotation is detected either by optical or magnetic sensors and can be directly related to the amount of slip experienced by the gripped object.
2. A probe is pressed against the surface of the gripped object and as it slips surface irregularities cause the probe to vibrate. Probe vibration caused by slip is detected

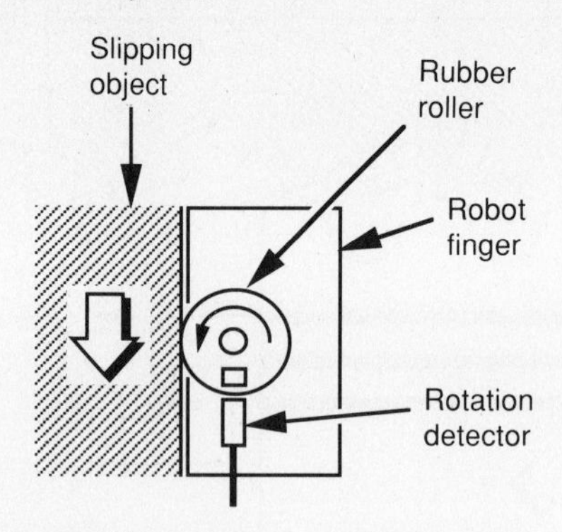


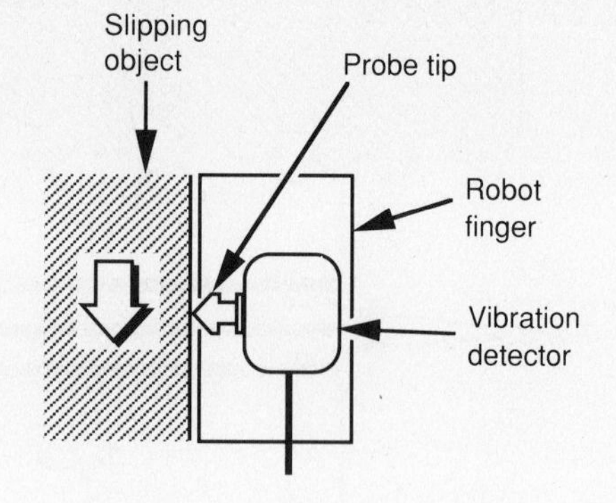


**Figure 7.14** Sensors to determine slip
(Redrawn from Ueda et al. 1972)

by a magnetic or piezoelectric transducer — very much like the pick-up of a gramophone.

## 7.5 Electrical conductivity

Electrical resistance can be used to monitor the contact between a robot end effector and an electrically conducting object. This section outlines two applications of electrical contact resistance sensing.

### 7.5.1 Control of deburring using electrical resistance

The automatic deburring of castings can be controlled by a hybrid force-position controller. This will ensure that the removal of metal takes place at a controlled rate; however, flatness of the resulting surface is hard to achieve because of variation between castings and wear of the grind wheel. A solution is to monitor the width of the region of the grind wheel touching the casting (Noda et al. 1985). One method of doing this is to set a radial conductive segment into the grind wheel surface (Figure 7.15). As the wheel rotates the proportion of time during which there is contact between the conductive line and object indicates the grind wheel width touching metal. When grinding the burr there is only a small region of contact between the contact and metal but as the burr is ground away the region of contact increases. Thus the size of the region of contact can be used to monitor the grinding process.

### 7.5.2 Resistance contact sensing in automatic sheep shearing

It is impossible to predict the precise body shape and posture of sheep during sheep shearing. Therefore, sensing is required to maintain the posture and clearance of the cutter assembly (Figure 7.16).

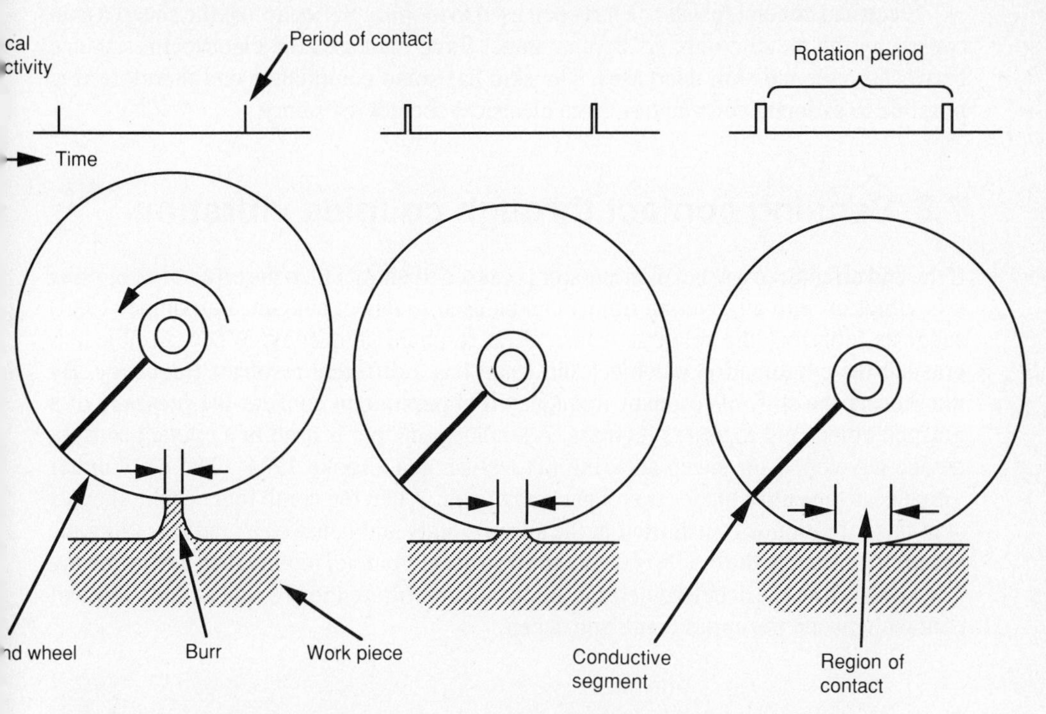


**Figure 7.15** Using electrical resistance to monitor deburring of metal castings (Adapted from Noda et al. 1985, with the permission of Prof. A. Noda)

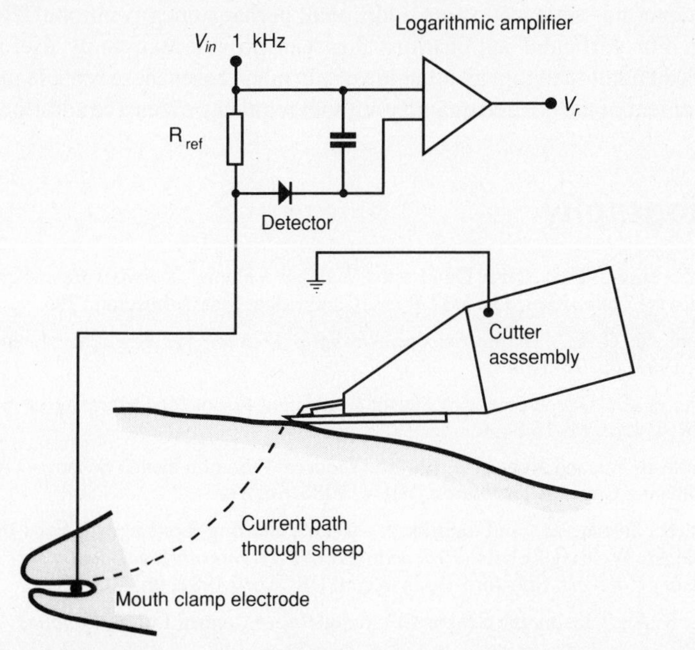


**Figure 7.16** Monitoring cutter clearance during sheep shearing using contact resistance (Redrawn from Trevelyan 1985, ©1985 MIT Press)

Electrical contact resistance has been used to monitor deflection of the sheep's skin caused by the metal comb. As comb contact force increases the electrical resistance between comb and skin decreases. The skin has some compliance and therefore it is possible to estimate deformation from electrical contact resistance.

## 7.6 Sensing contact through coupled vibration

If the end effector of a robot manipulator is caused to vibrate then the effects of coupling this vibration into an external object can be used to detect contact. Larcombe (1983) suggests vibrating the robot structure at its resonant frequency. When an object is grasped the combination of object and robot has a different resonant frequency. By monitoring the shift of resonant frequency it is possible to confirm the presence of a grasped object and to assess its mass. A similar principle is used in a contact-sensing device developed for sheep shearing (Trevelyan and Crooke 1984). The wool cutter vibrates as the cutter blades travel back and forth. When the comb touches the sheep's skin this vibration is transmitted to the sheep's body and hence to a cradle holding the sheep. Vibration of the cradle is detected by an accelerometer mounted on a tuned beam. The signal from the accelerometer is amplified and filtered to provide an indication of contact between the cutter comb and sheep.

## 7.7 Conclusion

This chapter has reviewed some additional, perhaps unconventional, forms of tactile sensing. For particular applications they can provide extremely useful information which is difficult to obtain by other means. In many cases these sensors produce a direct measurement of the desired quantity without requiring extensive additional processing.

## Bibliography

Hirose, S., et al., 'Titan III: A Quadruped Walking Vehicle', *Robotics Research — The Second International Symposium*, The MIT Press, Cambridge, Massachusetts, 1985.

Larcombe, M. H. E., 'Tactile Sensing', *Robotic Technology*, A. Pugh, ed., Peter Peregrinus, London, 1983, pp. 97–102.

Noda, A., et al., 'Development of Sensor Controlled Robot for Deburring', *Proceedings of the 15th ISIR*, Tokyo, 11–13 September 1985, pp. 207–14.

Patterson, R. W., and Nevill, G. E., 'The Induced Vibration Touch Sensor — A New Dynamic Touch Sensing Concept', *Robotica*, Vol. 4, 1986, pp. 27–31.

Presern, S., Spegel, M., and Ozimek, I., 'Tactile Sensing System with Feed-back Control for Industrial Arc Welding Robots', *Proceedings of the 1st International Conference on Robot Vision and Sensory Controls*, Stratford-Upon-Avon, UK, April 1981, pp. 205–13.

Russell, R. A., 'Closing the Sensor-Computer-Robot Control Loop', *Robotics Age*, April 1984, pp. 15–20.

Russell, R. A., 'Object Recognition Using Articulated Whisker Probes', *Proceedings of the 15th ISIR*, Tokyo, 1985, pp. 605–11.

Russell, R. A., 'Thermal Sensor for Object Shape and Material Constitution', *Robotica,* Vol. 6, 1988, pp. 31–4.

Trevelyan, J. P., 'Skills for a Shearing Robot: Dexterity and Sensing', *The Second International Symposium on Robotics Research*, H. Hanafusa and H. Inoue, eds, MIT Press, Cambridge, Massachusetts, 1985, pp. 273–80.

Trevelyan, J. P., and Crooke, M. D., 'Contact Sensing Device', Australian Patent No. 31508/84, 1984.

Ueda, M., et al., 'Tactile Sensors for Industrial Robot to Detect Slip', *Proceedings of the 2nd ISIR*, IIT Research Institute, Chicago, May 1972, pp. 63–76.

Wang, S. S. M., and Will, P. M., 'Sensors for Computer Controlled Mechanical Assembly', *The Industrial Robot*, March 1978, pp. 9–18.

## Questions

7.1 Describe the way in which a cat uses its vibrissae as a means of sensing its environment.

7.2 Why do you think that a cat's whiskers are distinctly curved rather than straight?

7.3 A high resonant frequency is desirable for a whisker probe. What is the reason for this and which design parameters control a whisker's resonant frequency?

7.4 Design a search strategy for:

(a) finding a randomly placed object on a table top; and
(b) tracing the outline of the object;

using a robot-mounted whisker probe which can only register contact using a simple switch. Consider the limitations of a whisker sensor for these applications.

7.5 Explain how different materials can be distinguished, based on their thermal properties.

7.6 An array thermal sensor can also function as a contact sensor to give a binary image of the area of contact between sensor and object. The binary image would be made up of those areas of the sensor which suffered a significant temperature drop on contact with the object. Explain the advantages and disadvantages of a thermal sensor for this application, compared to an imaging force sensor.

7.7 Describe a special purpose sensor for detecting the slippage of an object held in a robot gripper.

7.8 Describe a tactile sensor which uses electrical conductivity to detect contact and explain how the sensor can be used in a task requiring sensory feedback.

# Tactile sensing and mobility

Physical movement is almost essential for any form of tactile sensing. Unless objects can be constrained to bump into the sensor, the only way contact can be achieved is by giving the sensor some form of mobility and have it actively search for objects. For humans, hands are the organs of touch sensing. Equipped with their exquisitely sensitive skin, articulated fingers and kinesthetic senses, our hands provide us with a major source of information about our surroundings. When shown an unfamiliar object, we say, 'Let me see that,' but one of our instinctive reactions is to try to touch. For this reason, in many museums and art galleries we see notices proclaiming:

**'DO NOT TOUCH!'**

Robot grippers and tactile sensors are indispensable components of highly dextrous robot manipulator systems. Therefore, it is appropriate to devote some space, in a book on robot tactile sensing, to the subject of robot hand design. This chapter outlines some of the requirements of a robot hand and then provides a brief description of two dextrous hand designs and one dextrous gripper. In this context 'dextrous' indicates that:

1. it is possible to grasp a wide range of objects having different shapes and sizes; and
2. the hand/gripper can make small adjustments to the position and orientation of a grasped object without moving the rest of the robot.

## 8.1 Robot hands for gripping

The most obvious use for a robot hand is to grip things, and therefore robot hand design emphasizes this function. When gripping an object the goal is to control the object's position and orientation. A completely unconstrained object can move in six independent ways — translation in three orthogonal directions and rotation about these three orthogonal directions. We say the object has six degrees of freedom (DOF) (Figure 8.1).

As soon as an object comes into contact with another, its freedom of movement is reduced. The diagrams shown in Figure 8.2 provide some examples. To hold an object securely, its degrees of freedom must be reduced to zero. If friction is present, extra constraint is produced which reduces DOF and therefore it is worthwhile providing a high friction gripping surface for the fingers of a robot hand.

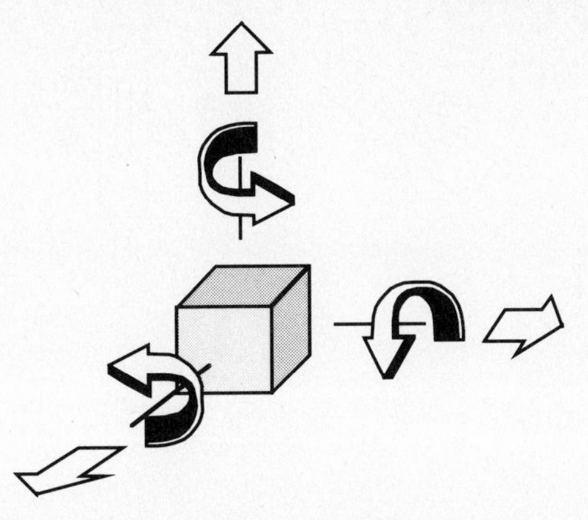

**Figure 8.1** The six degrees of freedom of an unconstrained object

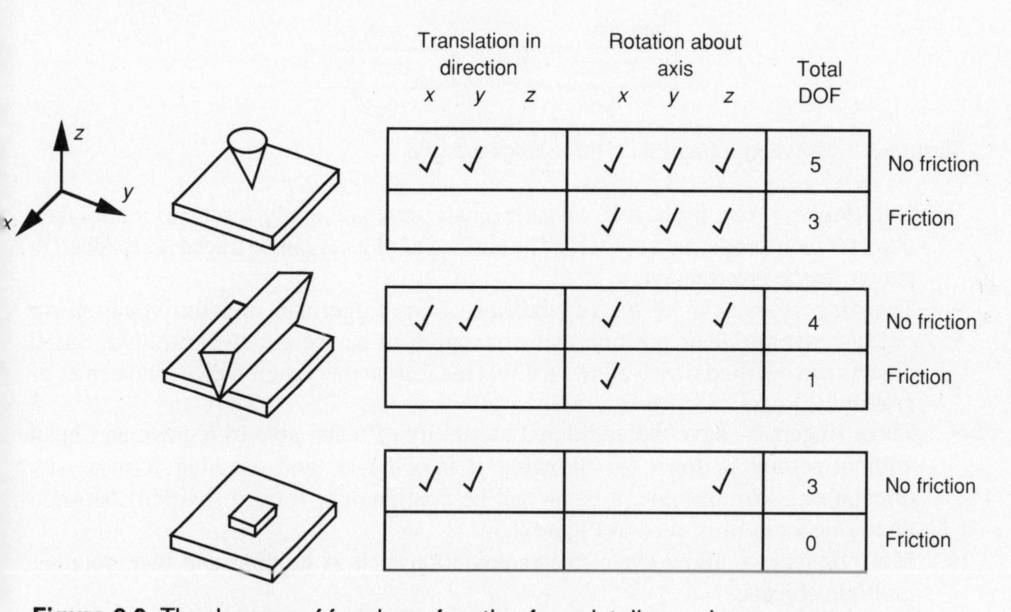

| Contact | Translation in direction $x$ | $y$ | $z$ | Rotation about axis $x$ | $y$ | $z$ | Total DOF | |
|---|---|---|---|---|---|---|---|---|
| Point | ✓ | ✓ | | ✓ | ✓ | ✓ | 5 | No friction |
| | | | | ✓ | ✓ | ✓ | 3 | Friction |
| Line | ✓ | ✓ | | ✓ | | ✓ | 4 | No friction |
| | | | | ✓ | | | 1 | Friction |
| Area | ✓ | ✓ | | | | ✓ | 3 | No friction |
| | | | | | | | 0 | Friction |



**Figure 8.2** The degrees of freedom of motion for point, line and area contact

## 8.2 How many fingers?

From the point of view of expense, complexity and reliability, it would be reasonable to design a robot hand with the minimum number of fingers necessary for its intended application. Here are some of the tasks which can be performed with a restricted number of fingers:

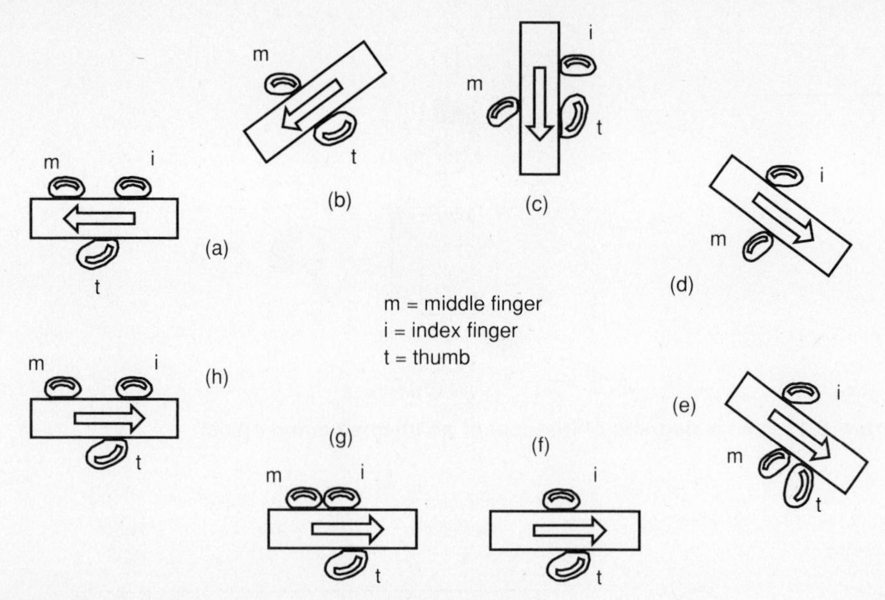


**Figure 8.3** Rotating a baton in a three-fingered grip

- One finger — can push, roll, or slide small parts and apply forces to manipulate objects (such as operate a switch). The surface of objects can be traced and probed to gather tactile information.
- Two fingers — have all the capabilities of one finger and in addition can grasp objects enabling their position and orientation to be accurately controlled. Small objects can be lifted which allows a direct measurement of their weight and center of gravity.
- Three fingers — have the additional capability of being able to regrasp an object without setting it down (or throwing it into the air and catching it in a new orientation). For example, a baton can be continuously rotated (twirled) between three fingers as illustrated in Figure 8.3.
- More fingers — allow even greater dexterity such as holding and manipulating multiple objects.

Having the ability to grasp objects between one or more fingers and a static palm, increases the dexterity of a hand. For example, a one-fingered hand could lift an object by gripping it between finger and palm. In effect, the palm is acting like an additional finger. Another strategy which increases the stability of grip is to mold the hand around an object producing multiple areas of contact between each finger and the grasped object.

## 8.3 The human hand

When grasping objects the human hand can be organized into a number of different configurations. Having five fingers with over twenty-five degrees of freedom, the human hand can exert a powerful grip and still deftly manipulate objects. Three of the many hand configurations we use for gripping are shown in Figure 8.4.

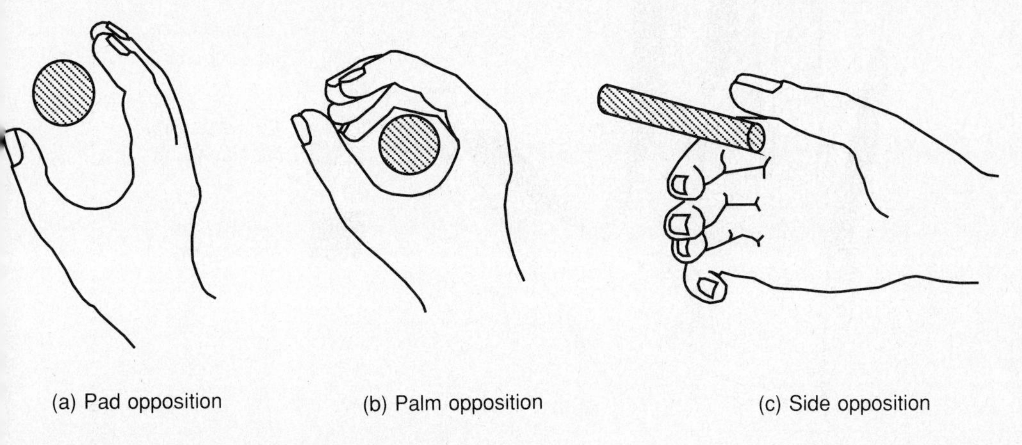


**Figure 8.4** Three hand configurations for gripping
(Redrawn from Iberall 1987, ©1987 IEEE)

1. Pad opposition gives:
   (a) finely controlled motion;
   (b) contact with the densest concentration of tactile sensors on the hand;
   (c) relatively poor stability; and
   (d) relatively poor maximum force.
2. Palm opposition is characterized by:
   (a) improved stability;
   (b) greater forces available from proximal finger joints;
   (c) little flexibility; and
   (d) reduced sensing.
3. Side opposition provides:
   (a) a compromise between flexibility, stability, and gripping force.

These basic grasps are modified to provide the appropriate blend of flexibility, stability, and gripping force required for a particular task.

Hand surgeons have found that it is important to maintain the skin sensation of certain areas of the hand, otherwise the dexterity of the hand is greatly reduced. Figure 8.5 illustrates those areas of the hand where skin sensation has been found to be important for gripping objects. This kind of information should be taken into account when positioning areas of sensory skin on a robot hand. In addition, a dextrous robot hand will also require a kinesthetic sense to determine the position of each finger joint and the forces applied by the fingers.

## 8.4 Dextrous robot hands and grippers

Many factors must be considered in the design of a dextrous robot manipulator, including (Kato 1982):

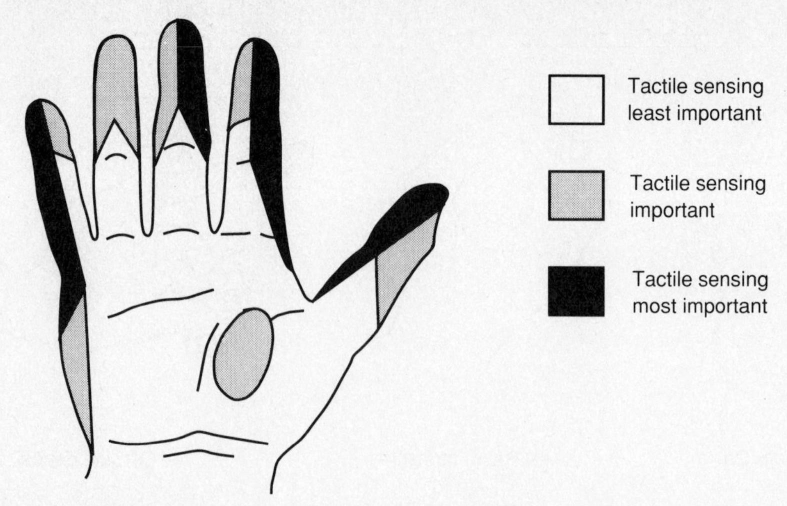


**Figure 8.5** The relative importance of touch sensation from different areas of the hand (Redrawn from Tubiana 1981, courtesy W. B. Saunders Co.)

- hand kinematics — number of fingers, type, and number of articulated joints in each finger, positioning of the fingers and lengths of each of its phalanges. These factors influence the dexterity of the hand or gripper;
- finger joint actuators — DC electric motors, pneumatic and hydraulic cylinders, shape-memory alloy (Nakano, Fujie, and Hosada 1984), and piezoelectric (Umetani and Suzuki 1980) actuators. The type of actuator determines the speed and strength of the hand/gripper;
- drive chain components — gears, rotating shafts, push rods, pulleys, or tendons (Rivin 1988); and
- number and location of sensors which affect the sensitivity and control of the hand/gripper.

In addition, the bearings and construction materials for the hand must be chosen.

## 8.5 Some examples of robot hand/gripper design

### 8.5.1 The Stanford–JPL dextrous hand

The Stanford-JPL hand (Salisbury and Craig 1982) has been designed with the least number of articulations and fingers which can still provide a six DOF incremental motion for a gripped object. If it is assumed that there is point contact with friction between each finger and object then three fingers are necessary to fully constrain the gripped object. To position and orientate the object each fingertip must be positioned accordingly. However, because there is only point contact the fingertip orientation does not matter (within certain limits) and therefore each finger only requires three DOF. The Stanford–JPL hand has been designed to maximize the dextrous volume of the fingers

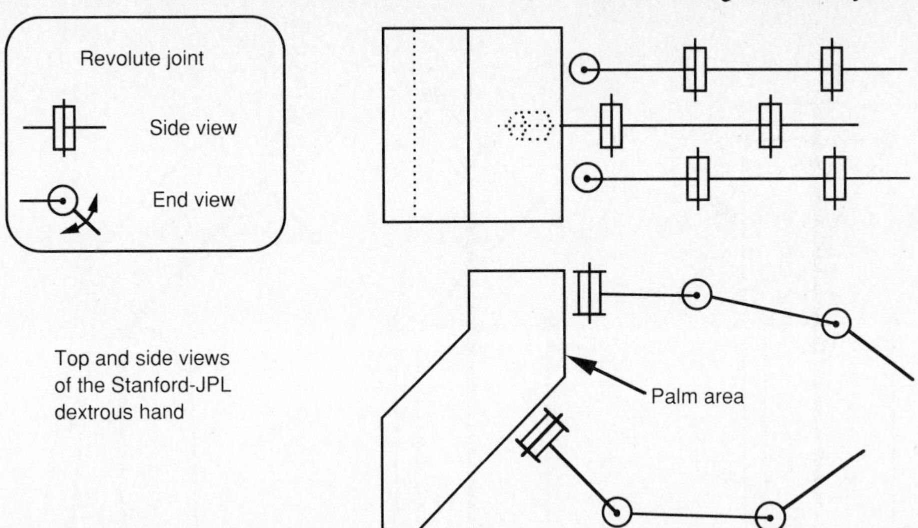


**Figure 8.6** Schematic diagram of the Stanford-JPL dextrous hand
(Redrawn from Salisbury and Craig 1982, ©1982 MIT Press)

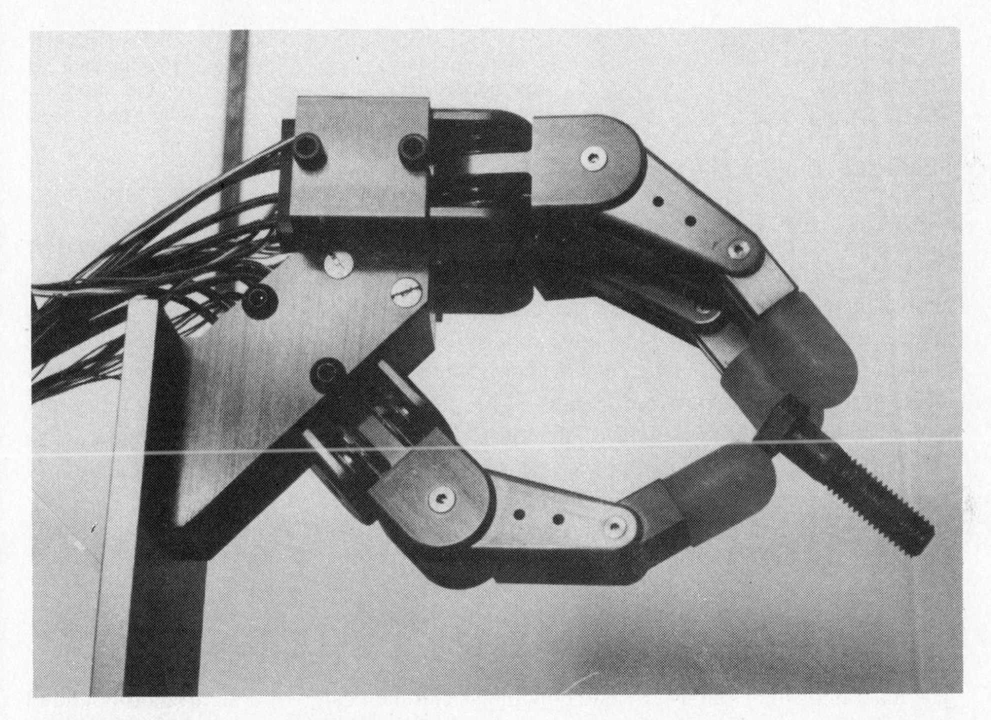

**Figure 8.7** The Stanford–JPL dextrous hand
(Photograph courtesy of David Lampe — MIT)

and also provide a palm so that the palm-opposition grasp can be utilized (see Figures 8.6 and 8.7).

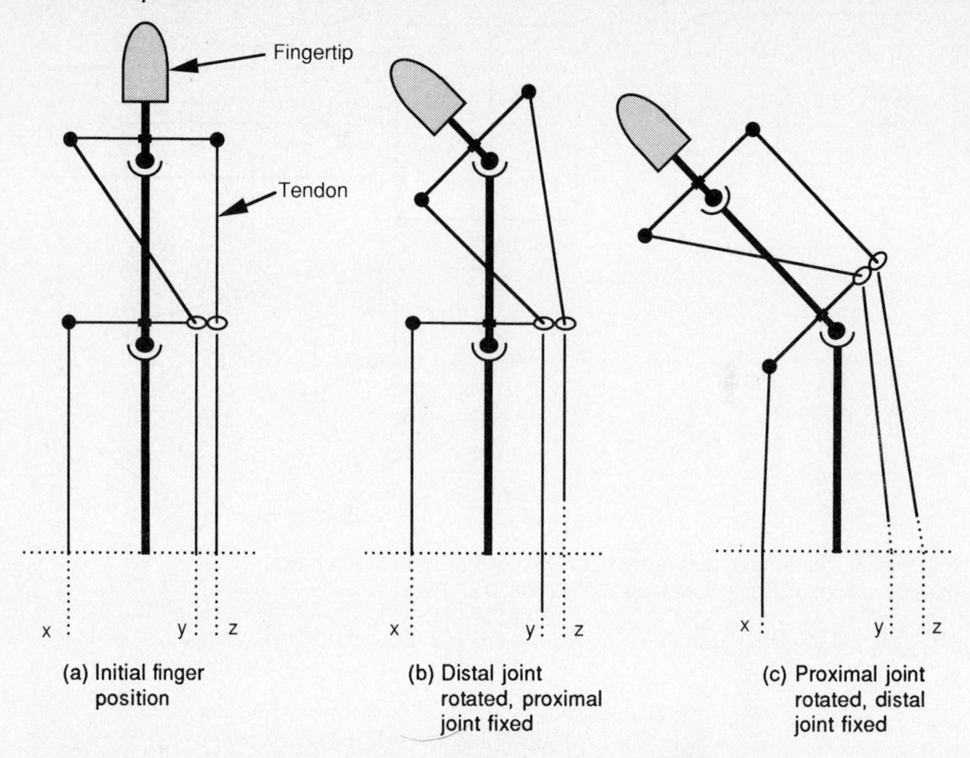


**Figure 8.8** The use of $n+1$ tendons to actuate an $n$-jointed finger

Each finger of this hand has the same design with three revolute joints actuated by teflon-coated tendons. Four DC servo motors are used to actuate each finger giving control of joint positions and also the tension in the tendons. The use of $n+1$ tendons to actuate $n$ joints is illustrated in Figure 8.8. In the case of finger joints, proximal means nearer to the wrist and distal means further from the wrist. The distal joint is actuated by a pair of tendons. As shown in Figure 8.8(b), pulling/releasing tendon y and releasing/pulling a corresponding length of tendon z will cause the distal joint to move. The proximal joint is adjusted by pulling/releasing tendon x and releasing/pulling tendons y and z an equal amount as illustrated in Figure 8.8(c).

The Stanford–JPL hand guides the tendons past each joint using pulleys. For simplicity the pulleys have been omitted from Figure 8.8.

## 8.5.2 The Utah–MIT dextrous hand

The makers of the Utah–MIT hand concluded that the design of a dextrous robot hand 'from the ground up' was too difficult (Jacobsen et al. 1986). They chose, instead, to base their design on the human hand. This biological model provides proof that such a design is capable of very intricate manipulation. Being closely modeled on the human hand it can also be used as a slave device in a teleoperation system (see Figures 8.9 and 8.10). Some changes have been made to the hand geometry to make it easier to build — the little

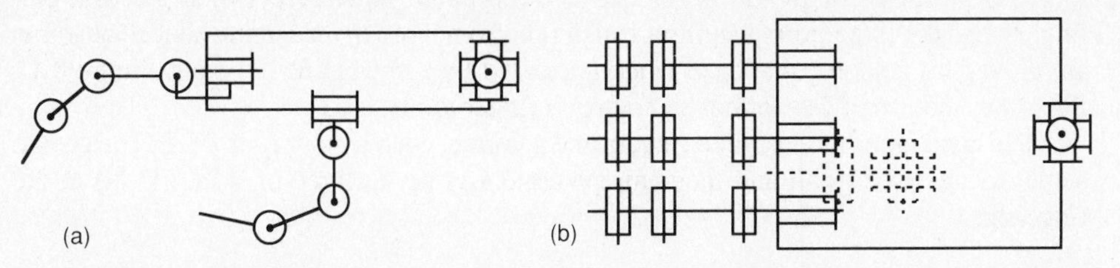


**Figure 8.9** A schematic diagram of the Utah-MIT dextrous hand: (a) top view, (b) side view (Redrawn from Jacobsen et al. 1986, © 1986 IEEE)

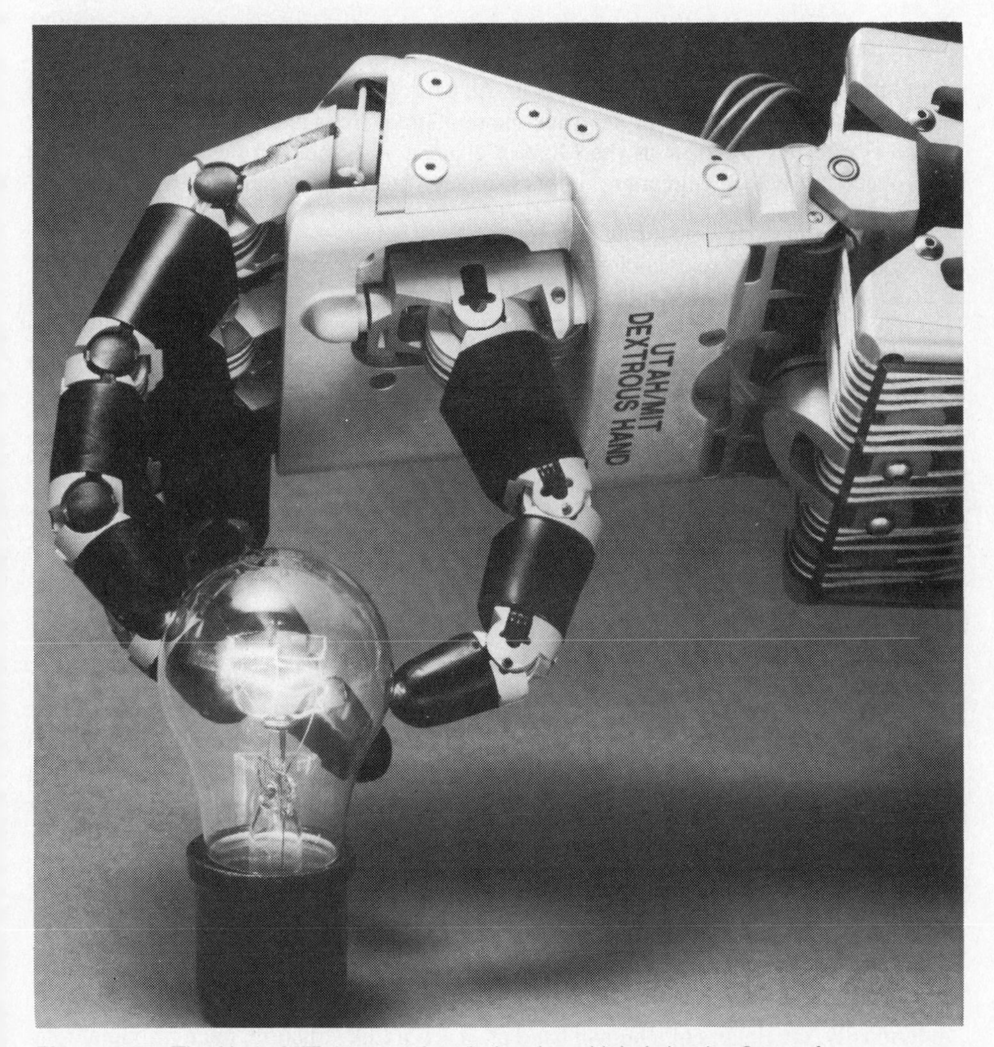


**Figure 8.10** The Utah–MIT dextrous hand, developed jointly by the Center for Engineering Design at the University of Utah and the Artificial Intelligence Laboratory at the Massachusetts Institute of Technology
(Photograph courtesy of Ed Rosenberger)

finger is omitted, the thumb is relocated to the palm, midway between the first two fingers, and the finger joint lengths are different from those in the human hand. The joints are moved by tendon pairs, each tendon actuated by a pneumatic cylinder. Figure 8.11 illustrates the use of $2n$ tendons to control $n$ finger joints.

The hand incorporates three fingers and a thumb, each having four DOF. Thirty-two actuators are used in all and they can generate a gripping force of 7 lb. (31 N) at the fingertips.

## 8.5.3 The Monash dextrous gripper

Robot end effectors that have a similar form to the human hand are called robot hands. Robotic devices for holding and manipulating objects which have a less anthropomorphic form are usually termed grippers. As part of a project to investigate the use of tactile sensory information in dextrous manipulation, the author designed the gripper which is referred to below as the Monash gripper. A number of design criteria were formulated to guide development of this gripper. These requirements were:

- adequate space for mounting sensors;
- good access for sensor cables;
- simple to construct and modify;
- dextrous — able to grasp a wide range of object shapes and also capable of imparting an incremental change in position and orientation. It has been estimated that a suitable range of incremental motion should be at least $\pm 1$ cm and $\pm 30^{\circ}$ (Salisbury and Craig 1982);
- easy to control; and
- light enough to be carried by a PUMA robot.

Taking into account these criteria, the structure of the gripper evolved from the following consideration:

### *8.5.3.1 Direct actuation*

In the human hand and the two examples of robot hands described earlier in this chapter, actuators are mounted remotely and control the fingers via tendons or cables. This removes the bulk and mass of the actuators to a more convenient location. The diagram shown in Figure 8.12 illustrates the probable effect of locating all the finger muscles in the palm of the hand. The palm is no longer hollow and can only grasp much smaller objects.

The effect of mounting the finger actuators closer to the fingers is to greatly simplify the actuator mechanism, making it easier to design, construct and control. For these reasons the actuators of the Monash dextrous gripper directly drive the fingers.

### *8.5.3.2 Parallel link 'finger' structure*

Individually, the fingers of a human hand are a serial link structure. Each proximal joint supports the whole of the more distal part of the finger. This is illustrated, for a simple two-joint finger, in Figure 8.13(a). An alternative organization is to have all the finger joints in parallel, as shown in Figure 8.13(b).

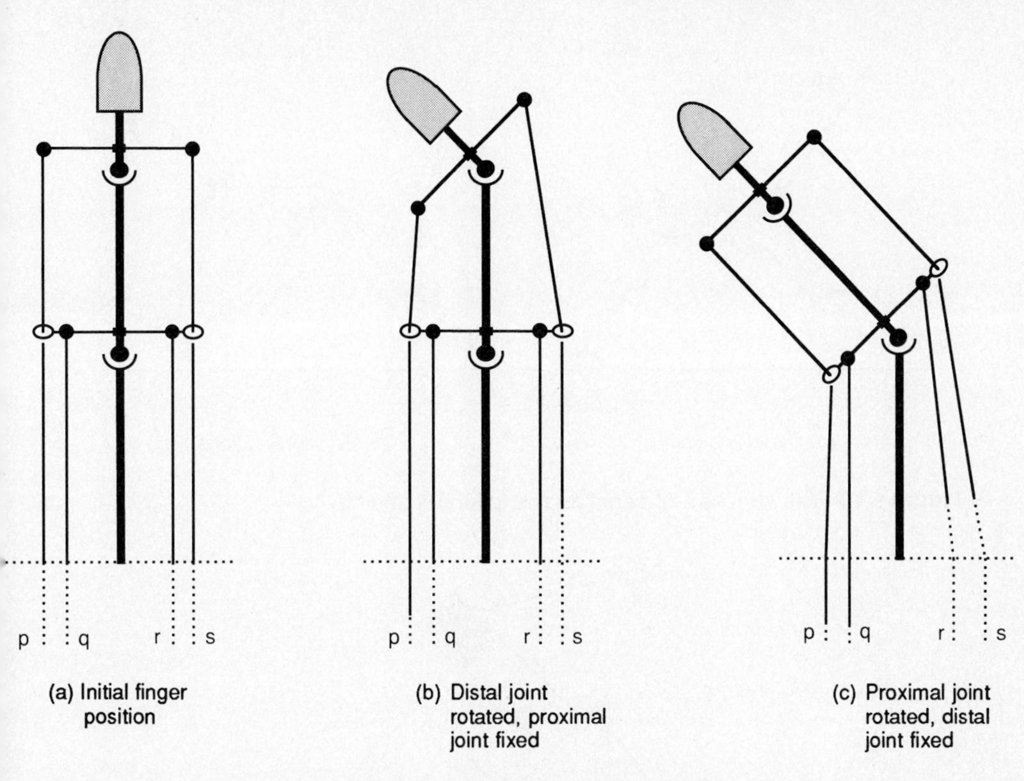


(a) Initial finger position

(b) Distal joint rotated, proximal joint fixed

(c) Proximal joint rotated, distal joint fixed

**Figure 8.11** The use of $2n$ tendons to actuate an $n$-jointed finger

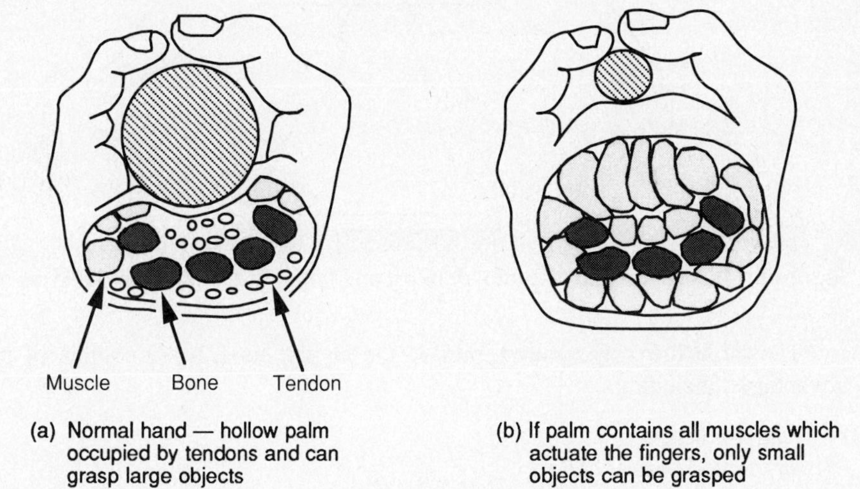


(a) Normal hand — hollow palm occupied by tendons and can grasp large objects

(b) If palm contains all muscles which actuate the fingers, only small objects can be grasped

**Figure 8.12** The consequences of actuating the fingers via tendons or directly
(Redrawn from Kapandji 1981, courtesy W. B. Saunders Co.)

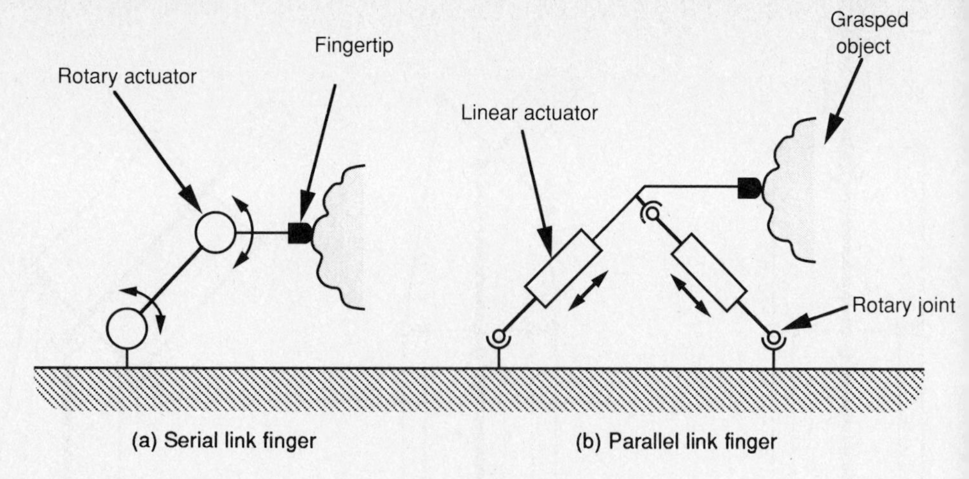


**Figure 8.13** The structure of serial and parallel link fingers

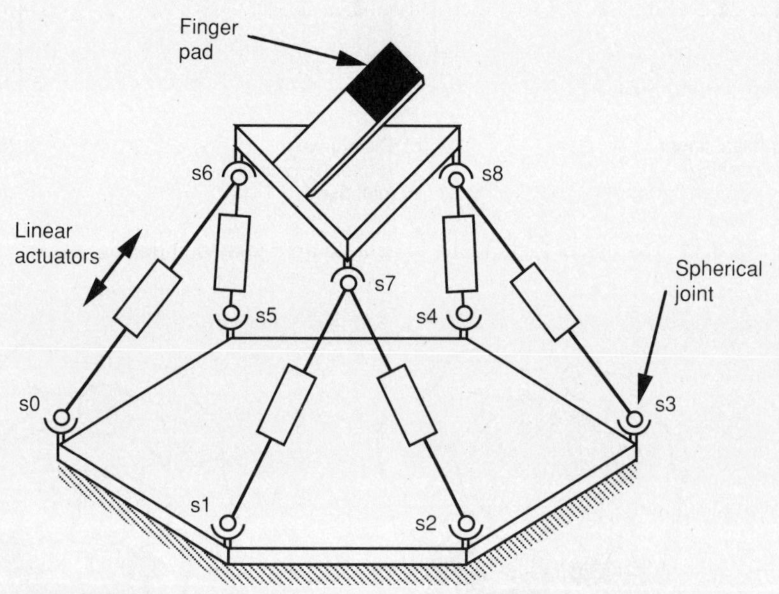


**Figure 8.14** A schematic diagram showing one finger of the Monash dextrous gripper

The serial link organization, seen in the human hand, has a number of potential advantages, including:

- longer reach;
- larger work space; and
- each finger can provide multiple points of contact (by wrapping around the gripped object), thus improving grip stability.

By comparison, there are also points in favor of the parallel link structure, including:

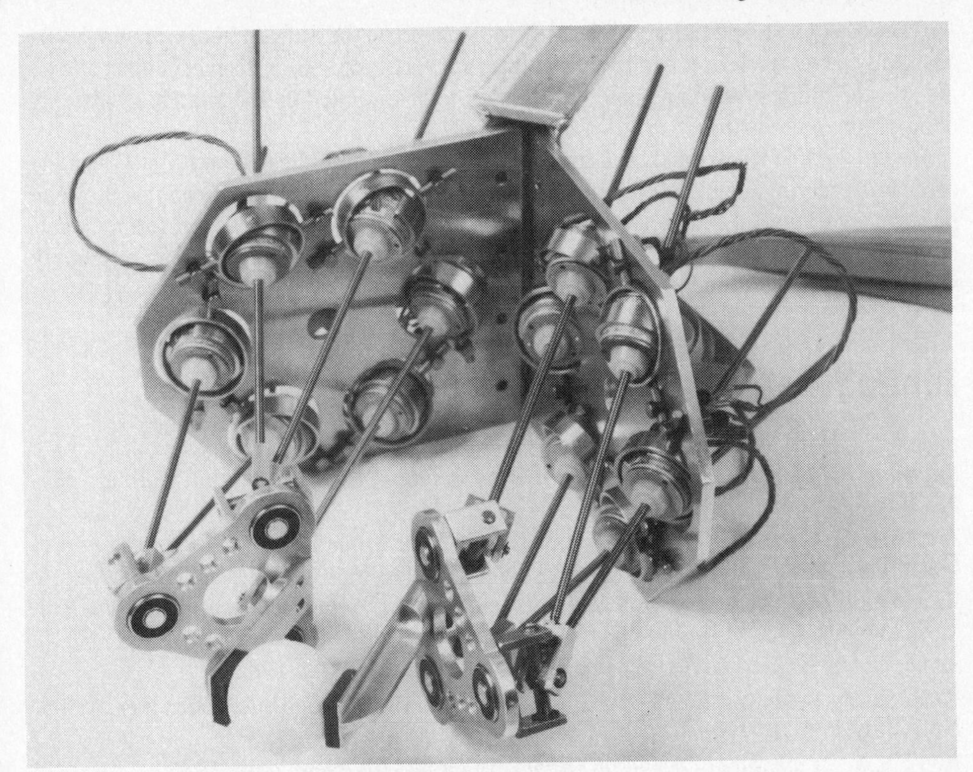

**Figure 8.15** The Monash dextrous gripper

- it is more rigid;
- errors in one actuator do not propagate through the whole structure;
- it can incorporate a large conduit for sensor wires;
- it allows all actuators to be mounted at the finger base without indirect linkages; and
- a 6 DOF finger is readily obtained.

On balance, it was decided that the parallel link structure best fitted the original design criteria.

### *8.5.3.3 Two fingers*

The Monash gripper employs two fingers, each having six DOF. In order to hold an object securely, planar contact with friction will be required between each finger and the gripped object. In the design of mechanical hands it is common to model the contact between finger and grasped object as a kinematic pair (Salisbury and Roth 1983). Thus, it is assumed that there will be relative motion between finger and object. The Monash gripper is provided with compliant fingertips and it is the expectation that these will provide a high degree of kinematic coupling between finger and object, thus immobilizing the object in the grasp. The six DOF of each finger will enable the gripped object to be repositioned without changing the spatial relationship between fingers and object. This greatly simplifies control of the gripper.

The structure of each finger (Figure 8.14) is based on the Stewart Platform which was originally developed for use in flight trainers (Stewart 1965–66). This mechanism allows the six step motors, which actuate each finger, to be mounted on the hand base plate. A photograph of the gripper is shown in Figure 8.15.

Multi-jointed robot hands and grippers can be more complicated and costly than the robot arm carrying them. Manufacturing industry will need to be convinced that they can perform a very useful function before investing in such devices. However, future developments which combine sensory feedback and robot grippers will create an extremely versatile sensing and manipulation device.

## Bibliography

IBERALL, T., 'The Nature of Human Prehension: Three Dextrous Hands in One', *IEEE International Conference on Robotics and Automation*, Raleigh, North Carolina, 1987, pp. 396–401.

JACOBSEN, S. C., et al., 'Design of The Utah/MIT Dextrous Hand', *IEEE International Conference on Robotics and Automation*, San Francisco, California, 1986, pp. 1520–32.

KAPANDJI, I. A., 'The Upper Limb as Logistical Support for the Hand', in *The Hand*, Vol. 1, R. Tubiana, ed., W. B. Saunders Co., Philadelphia, 1981.

KATO, I., *Mechanical Hands Illustrated*, Survey Japan, Tokyo, 1982.

NAKANO, Y., FUJIE, M., and HOSADA, Y., 'Hitachi's Robot Hand', *Robotics Age*, Vol.6, No.7, July 1984, pp. 18–20.

RIVIN, E. I., *Mechanical Design of Robots*, McGraw-Hill, New York, 1988.

SALISBURY, J. K., and CRAIG, J. J., 'Articulated Hands: Force Control and Kinematic Issues', *The International Journal of Robotics Research*, Vol. 1, No. 1, Spring 1982, pp. 4–17.

SALISBURY, J. K., and ROTH, B., 'Kinematic and Force Analysis of Articulated Mechanical Hands', *Journal of Mechanisms, Transmissions, and Automation in Design*, Vol. 105, March 1983, pp. 35–41.

STEWART, D., 'A Platform with Six Degrees of Freedom', *Proceedings of the Institute of Mechanical Engineers*, Vol. 180, Part 1, No. 15, 1965–6, pp. 371–86.

TUBIANA, R., 'Architecture and Functions of the Hand', in *The Hand*, Vol. 1, R. Tubiana, ed., W. B. Saunders Co., Philadelphia, 1981.

UMETANI, Y., and SUZUKI, H., 'Piezo-Electric Micro-Manipulator in Multi-Degrees of Freedom with Tactile Sensibility', *Proceedings of the 10th ISIR*, Milan, Italy, 1980, pp. 571–9.

## Questions

8.1 What are the special capabilities of a dextrous robot hand/gripper?

8.2 Using qualitative reasoning, determine:

(a) the minimum number of fingers required to hold an object (but not necessarily prevent it rotating in the grip); and

(b) the minimum number of fingers required to hold an object *and* prevent its rotation (for some objects this may not be possible) for the following cases:

(i) the fingers make point contact without friction;
(ii) the fingers make point contact with friction;
(iii) the fingers make area contact without friction; and
(iv) the fingers make area contact with friction.

8.3 Referring to Figure 8.6, it appears that the tactile sensibility of a large area of the little finger is important for maintaining the dexterity of the human hand. Explain why you think this should be the case.

8.4 Compare the dexterity of the human hand with that exhibited by the following:

(a) the weaverbird — a small, seed-eating bird which weaves an intricate nest from grass and other plant material;
(b) the termite — a large social insect which builds complex and extensive mud nests; and
(c) the beaver — a rodent which cuts down trees and uses them to build dams across fast-flowing streams.

How is the dexterity achieved? Are there any lessons here for robot designers?

8.5 Briefly describe the main features of the Stanford-JPL dextrous hand.

8.6 Compare the advantages and disadvantages of serial/parallel actuation for the fingers of a dextrous hand/gripper.

8.7 Compare the advantages and disadvantages of:

(a) a dextrous robot hand which can be used for many sensing and manipulation operations; and
(b) a special purpose gripper which has been designed to perform one or a small number of specific tasks.

# 9 Tactile feedback at the reflex level

In the human body there are numerous feedback loops which tend to maintain the body in a steady state unless overridden by a higher level command.

The muscle stretch reflex (Figure 9.1) is one of these feedback loops which helps to maintain body posture. Involuntary lengthening of a muscle is detected by neuromuscular spindles and signaled to the spinal cord via afferent neurons. Providing the reflex is not being inhibited by signals from higher in the nervous system, the muscle is stimulated to contract via motor neurons, thus resisting the forces causing it to stretch. This action is controlled at a low level within the nervous system; no higher functions of reasoning or intelligence are involved.

In a similar manner, the actions of a robot can be controlled directly using the output of sensors with a minimum of additional processing. Therefore, if the use of tactile feedback within a robot system involves no high-level reasoning or intelligence it may be categorized as reflex control. Examples would be:

1. to maintain some quantity constant, such as:
   (a) the wool cutter posture during sheep shearing;
   (b) the applied force in burnishing and grinding; and
   (c) the tactile image on a finger pad — to detect and correct for slip; and

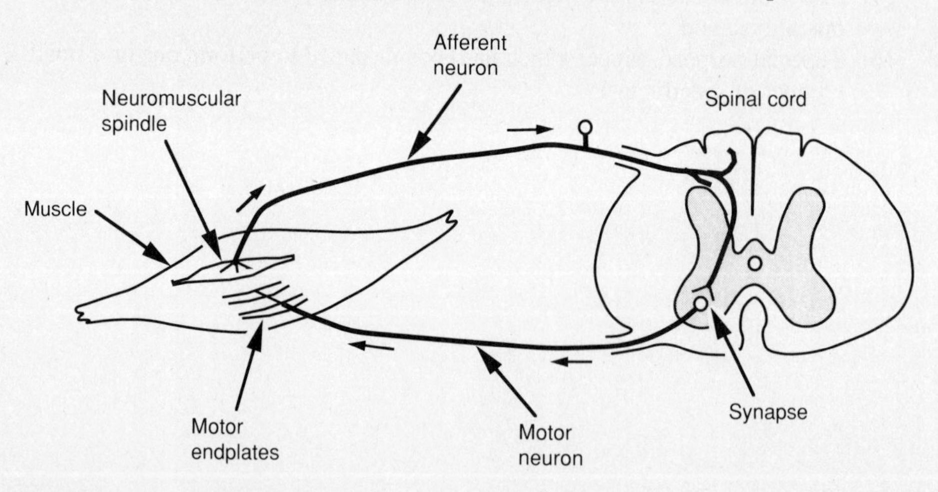


**Figure 9.1** The muscle stretch reflex

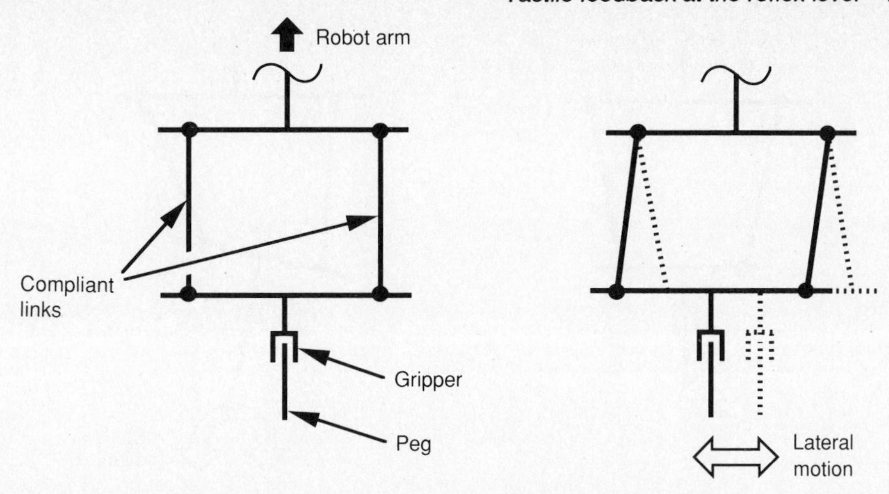


**Figure 9.2** Compliant structure to accommodate lateral misalignment

2. to optimize some quantity, such as:
   (a) minimizing reaction forces during insertion and other assembly operations; and
   (b) withdrawing a robot gripper when an excessive temperature is detected.

## 9.1 Selective compliance

A solution to many tracking and insertion problems can be formulated in terms of selectively altering the mechanical compliance of the robot structure. Selective compliance can be of benefit during assembly tasks, edge and surface following, and grinding operations.

### 9.1.1 Passive mechanical compliance

An excellent illustration is the remote center compliance (RCC) mechanism used to aid assembly (Whitney and Nevins 1986). If a peg is offered into a hole with no misalignment then the peg will slide in correctly. Lateral and rotational errors will tend to cause jamming, which can become worse if the peg is pushed from behind. If the peg could be pulled into the hole then jamming would not occur. The RCC alters the motion of the peg in response to jamming forces so that it reacts as though it were being pulled into the hole from in front rather than being pushed from behind. This type of motion is achieved by the following two compliance mechanisms.

Parallel compliant links, shown in Figure 9.2, allow side to side motion which accommodates lateral misalignment. Compliance of the links can be achieved, either by constructing the links of a flexible material, or by incorporating some rotational freedom into the joints at each end of the 'compliant' links.

Angled compliant links, shown in Figure 9.3, allow rotation about a point ahead of the peg and this accommodates angular misalignment of peg and hole.

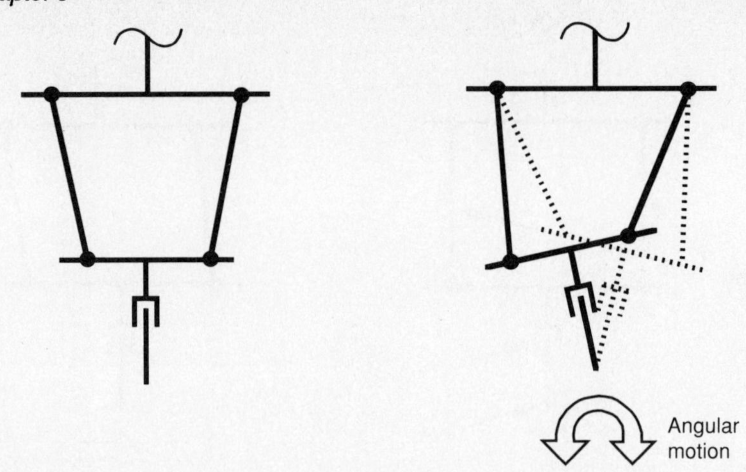


**Figure 9.3** Compliant structure to accommodate rotational misalignment

The RCC is a purely passive, mechanical device. A similar effect can be achieved by sensing the forces generated during insertion operations and controlling the robot joint servos in a suitable manner. At present RCC devices are faster and considerably cheaper than using servo control to achieve the same compliant motion. However, servo control is more adaptable in that the magnitude and direction of the robot compliance can be changed under program control.

## 9.1.2 Active control of compliance

There are two basic strategies for this kind of robot control — logic branching and continuous feedback (Whitney 1987). Continuous feedback is similar to conventional servo control and implements a sampled data control system. A generalized block diagram of such a system is shown in Figure 9.4.

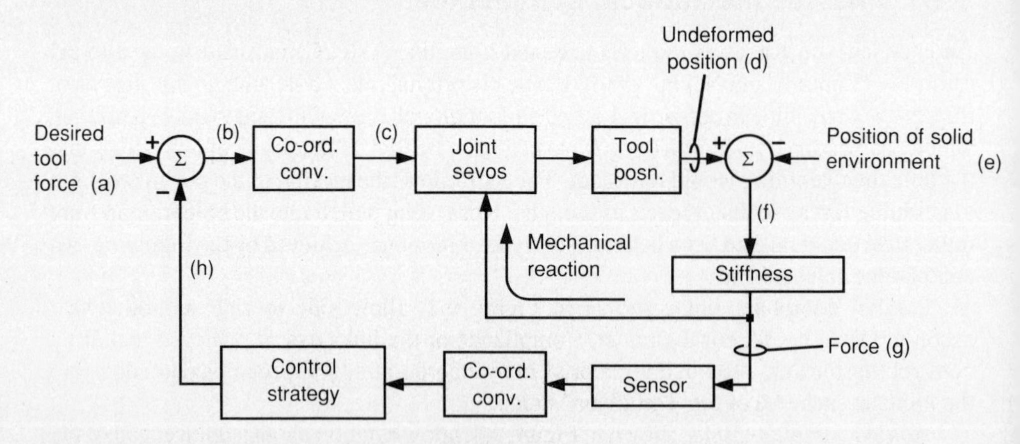


**Figure 9.4** Architecture of robot force feedback
(Adapted from Whitney 1987, ©1987 MIT Press)

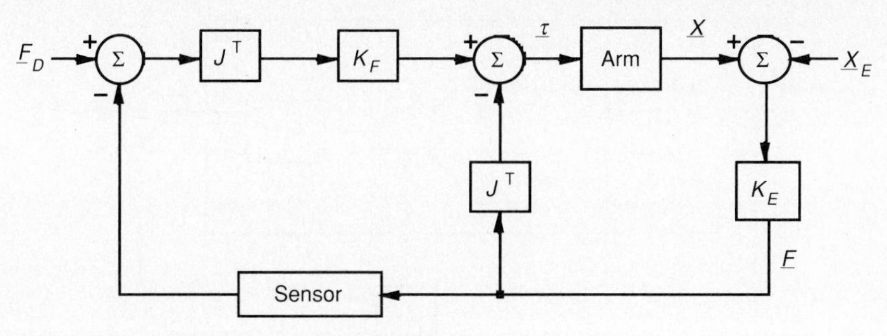


**Figure 9.5** Force control of a robot manipulator
(Redrawn from Whitney 1987, ©1987 MIT Press)

The flow of signals around this diagram can be summarized as follows. Feedback of applied force (h) is subtracted from the desired tool force (a) to give an error signal (b). The error signal (b) is in world coordinates and must be converted to joint coordinates (c) by a coordinate transformation. Servo motors try to drive the robot arm to a particular point in space (d) which for force control is also occupied by an external object (e). Because two objects (the tool and the external object) cannot both occupy the same point in space, the robot structure, sensors, tool, and the external object must all deflect a little (f) and, depending on the combined stiffness of all these objects, a force results (g). The force is registered by a sensor and processed to give the feedback signal (h).

Two examples taken from Whitney (1987) will be examined in more detail. The first example (Figure 9.5) maintains a constant force between the tool and environment (perhaps during a grinding operation, in which case the environment might be a casting and the tool a grind wheel).

The symbols used in Figure 9.5 are as follows:

Arm = the arm mechanics which convert servo torques (in joint coordinates) into tool position (in world coordinates);
$\underline{F}$ = the contact force between tool and environment;
$\underline{F}_D$ = the desired contact force;
$J$ = the arm Jacobian which relates small changes in the robot arm servo motor positions/angles to the resulting small changes in the tool position (measured in world coordinates). *Note:* the Jacobian varies with change in the robot arm configuration;
$K_F$ = an adjustable gain in the feedback loop;
$K_E$ = the total stiffness (force/distance) of all contacting items (arm, sensor, environment, etc.);
Sensor = a sensor that converts the forces generated by contact into a signal suitable for processing by the force control system;
$\underline{X}$ = the position of the tool if solid contact had not been made;
$\underline{X}_E$ = the position of contact with the solid environment; and
$\underline{\tau}$ = torques generated by the arm servos.

Force control only works while there is contact between tool and environment. If contact is lost then the robot arm can be subjected to large accelerations unless this condition is detected

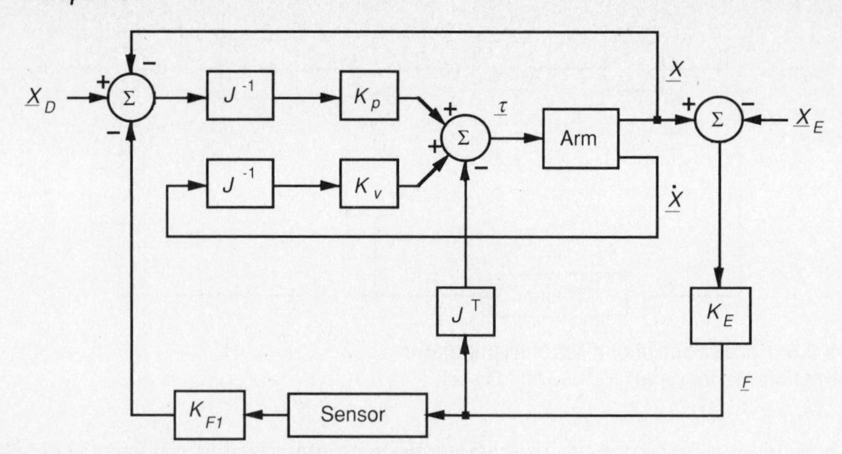


**Figure 9.6** Stiffness control of a robot manipulator
(Redrawn from Whitney 1987, ©1987 MIT Press)

and the force control disconnected. An alternative control strategy is to use a position controller but modify the apparent stiffness of the robot arm by applying force feedback as well. The implementation of stiffness control is illustrated by the block diagram shown in Figure 9.6. Two loops feed back position and velocity to provide position control. An additional loop feeds back force so that the robot will stop short of a desired position if a certain force is encountered.

The additional symbols used in Figure 9.6 are:

$\underline{X}_D$ = desired position;
$K_p$ = adjustable gain on feedback proportional to *position* error;
$K_V$ = adjustable gain on feedback proportional to *velocity*; and
$K_{F1}$ = matrix to adjust the apparent stiffness of the robot arm.

There are several other forms of force feedback including damping control, impedance control, and hybrid control (Whitney 1987) which also modify the compliance of the robot arm.

A similar effect can be obtained by formulating the control as a computer program. Thus, to insert a peg into a hole with the hole axis aligned in the *z* direction the following instructions could be used:

```
COMPLY FORCE X 0        {move to reduce forces in x direction to zero}
COMPLY FORCE Y 0        {move to reduce forces in y direction to zero}
COMPLY TORQUE X 0       {rotate to reduce torques about x axis to zero}
COMPLY TORQUE Y 0       {rotate to reduce torques about y axis to zero}
STOP FORCE Z -1         {stop when force in negative z direction ≥ 1 }
MOVE Z                  {move in the positive z direction}
```

(Program reproduced from Whitney 1987, ©1987 MIT Press.)

The COMPLY and STOP commands are obeyed while the following MOVE command is being executed and therefore the robot will move in the *z* direction while accommodating forces and torques in $x$ and $y$ directions. Both logic branching and continuous feedback can be used to implement force feedback. They both present very similar problems of system stability. When touching a surface with a large value of $K$ (stiffness) the robot arm must exhibit a low $K$ to achieve stability. Controlling a robot arm to give compliant motion leads to sluggish response. This can be counteracted by making the robot arm smaller, hence improving its dynamic response, or by incorporating passive compliance in the robot structure.

## 9.2 Detecting slip by tracking image center of area and axis of minimum moment of inertia

If the finger pads of a robot gripper incorporate force sensor arrays the output from the sensors may be used to detect slip. As an object slips through the fingers the sensor tactile image will change. Therefore, a simple strategy would be to monitor the tactile image and interpret any change as the result of slip. Unfortunately, the force image may change if the grip is tightened without this causing any movement of the gripped object. More information can be derived from a tactile image by using image processing techniques developed for analysis of binary visual images. These techniques can be used to determine speed and direction of slip and also whether the object is tending to rotate (indicating that the center of gravity of the object is not vertically in line with the finger pads).

Assume that the output of the tactile sensor is a two-dimensional array of values with $i$ rows and $j$ columns (Figure 9.7). Each touch sensitive point produces an output $t_{ij}$ which can have a value between $t_{min}$ and $t_{max}$.

The sensor output $t_{ij}$ is converted to a binary value $\tau_{ij}$ as follows:

$$\tau_{ij} = 0 \text{ if } t_{min} \leq t < t_{threshold}$$
$$\tau_{ij} = 1 \text{ if } t_{threshold} \leq t \leq t_{max}$$

Sensors with an output above the threshold are assumed to be in contact with the object and those below not in contact.

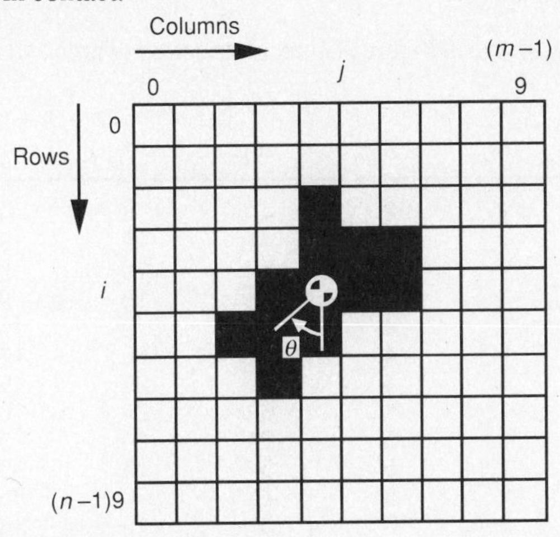


**Figure 9.7** A 10 x 10 binary tactile image

The technique uses some of the lower order moments of area of the image. Moments of area are defined as:

$$m_{pq} = \sum_i \sum_j i^p \cdot j^q \cdot \tau_{ij} \tag{9.1}$$

Thus the total area of contact is:

$$A = m_{00} = \sum_{i=0}^{n-1} \sum_{j=0}^{m-1} \tau_{ij} \tag{9.2}$$

A measure of the position of the tactile image can be found by determining the location of its center of area in both row and column coordinates:

$$i_0 = \frac{m_{10}}{m_{00}} = \frac{1}{A} \sum_{i=0}^{n-1} \sum_{j=0}^{m-1} i \cdot \tau_{ij} \tag{9.3}$$

$$j_0 = \frac{m_{01}}{m_{00}} = \frac{1}{A} \sum_{i=0}^{n-1} \sum_{j=0}^{m-1} j \cdot \tau_{ij} \tag{9.4}$$

Orientation can be estimated by tracking variation in the axis of minimum moment of inertia of the image:

$$\theta_0 = \frac{1}{2} \tan^{-1} \left[ \frac{2(m_{00} \cdot m_{11} - m_{10} \cdot m_{01})}{(m_{00} \cdot m_{20} - m_{10}^{\,2}) - (m_{00} \cdot m_{02} - m_{01}^{\,2})} \right] \tag{9.5}$$

There is an ambiguity in the calculation of the inverse tangent. However, the sign of both the numerator and denominator in Equation 9.5 are known and therefore the ambiguity can be resolved. Figure 9.8 shows the relationship between the sign of numerator and denominator and the corresponding range of $\theta_0$.

*Example*: Using the binary image in Figure 9.7 calculate center of area and axis of minimum moment of inertia.

| Pixel | $i$ | $j$ | $m_{00}$ | $m_{10}$ | $m_{01}$ | $m_{11}$ | $m_{20}$ | $m_{02}$ |
|---|---|---|---|---|---|---|---|---|
| 0 | 2 | 4 | 1 | 2 | 4 | 8 | 4 | 16 |
| 1 | 3 | 4 | 1 | 3 | 4 | 12 | 9 | 16 |
| 2 | 3 | 5 | 1 | 3 | 5 | 15 | 9 | 25 |
| 3 | 3 | 6 | 1 | 3 | 6 | 18 | 9 | 36 |
| 4 | 4 | 3 | 1 | 4 | 3 | 12 | 16 | 9 |
| 5 | 4 | 4 | 1 | 4 | 4 | 16 | 16 | 16 |
| 6 | 4 | 5 | 1 | 4 | 5 | 20 | 16 | 25 |
| 7 | 4 | 6 | 1 | 4 | 6 | 24 | 16 | 36 |
| 8 | 5 | 2 | 1 | 5 | 2 | 10 | 25 | 4 |
| 9 | 5 | 3 | 1 | 5 | 3 | 15 | 25 | 9 |
| 10 | 5 | 4 | 1 | 5 | 4 | 20 | 25 | 16 |
| 11 | 6 | 3 | 1 | 6 | 3 | 18 | 36 | 9 |
| | | | 12 | 48 | 49 | 188 | 206 | 217 |

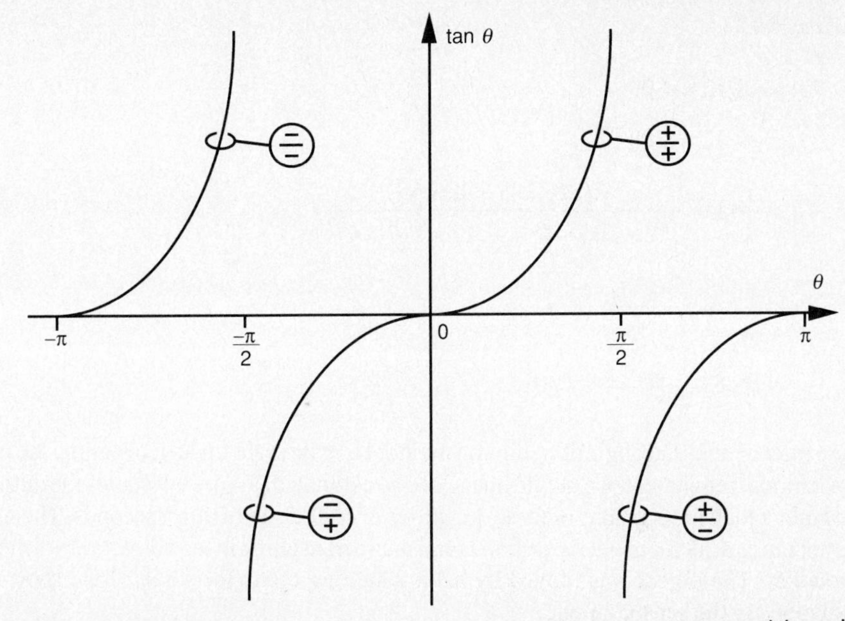


**Figure 9.8** Relationship between the sign of numerator and denominator and the value of inverse tangent

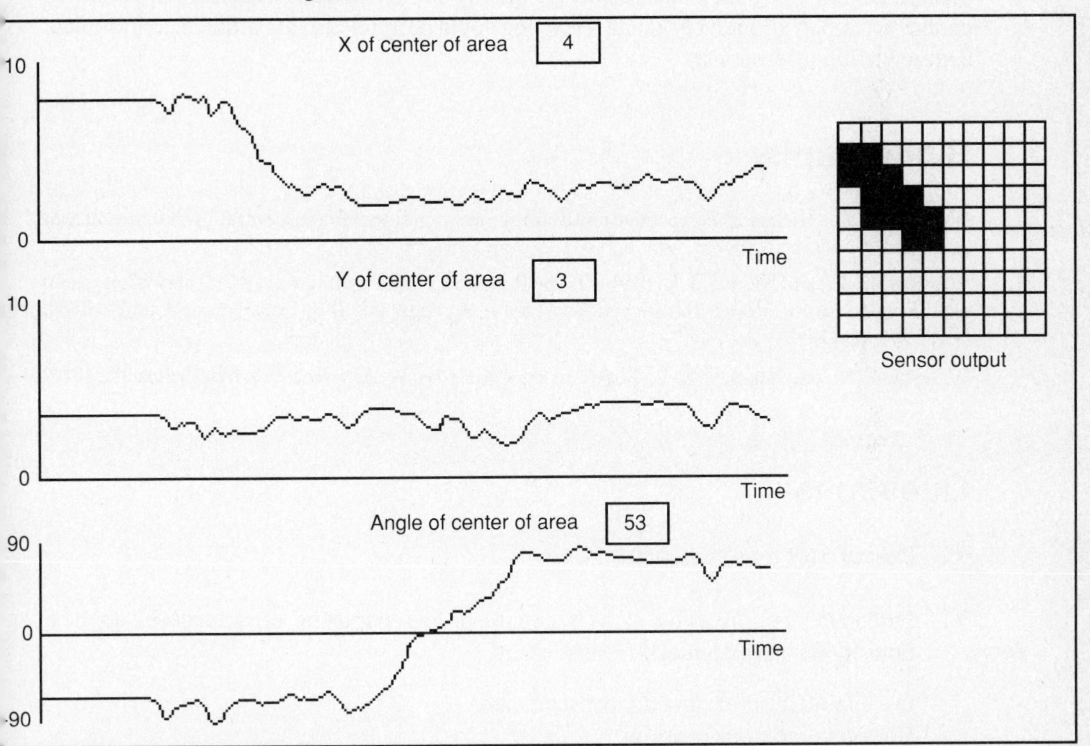


**Figure 9.9** Tracking movements of center of area and axis of minimum moment of inertia of an object moved over the surface of a 10 x 10 imaging force sensor

Area = 12

$$i_0 = 48/12 = 4.00$$

$$j_0 = 49/12 = 4.08$$

$$\theta_0 = \frac{1}{2}\tan^{-1}\frac{2(12\times188-48\times49)}{(12\times206-48^2)-(12\times217-49^2)}$$

$$= \frac{1}{2}\tan^{-1}\frac{-192}{-35}$$

$$= 39.83^\circ - 90^\circ = -50.16^\circ$$

If the center of area and angle of minimum moment of inertia are tracked over time the linear and rotational movement of a tactile image can be estimated. Figure 9.9 shows the output of a program which plots graphs of these quantities over a period of thirty seconds. The image does not move at all for about five seconds and then a translation in the $x$ direction is followed by rotation. The object was moved by hand which accounts for some of the short-term fluctuations in the sensor output.

Using this kind of information a purely reflex tightening of grip or rotation of the manipulator to bring the object center of gravity in vertical line with the gripper could counteract slip. Alternatively, some reasoning could be applied to the problem to formulate a more intelligent response.

## Bibliography

WHITNEY, D. E., 'Historical Perspective and State of the Art in Robot Force Control', *The International Journal of Robotics Research*, Vol. 6, No. 1, Spring 1987, pp. 3–14.

WHITNEY, D. E., and NEVINS, J. L., 'What Is the Remote Center Compliance (RCC) and What Can It Do?' *Robot Sensors*, Vol. 2, *Tactile and Non-Vision*, A. Pugh, ed., IFS (Publications) Ltd, Bedford, UK, 1986, pp. 3–15.

WINSTON, P. H., and HORN, B. K. P., *Lisp* (2nd Ed.), Addison-Wesley, Reading, Massachusetts, 1984.

## Questions

9.1 Describe the muscle stretch reflex.

9.2 Extend the lists presented at the beginning of this chapter of reflex feedback applications in industrial/domestic robotics that:

(a) maintain some quantity constant; and
(b) optimize some quantity.

9.3 How does a remote center compliance (RCC) work and what is it used for?

9.4 Why is it important to be able to detect the slipping of a grasped object?

9.5 Describe a control scheme to provide active control of robot stiffness.

9.6 Write a computer program to perform the following tasks:

(a) close a sliding door; and
(b) close a hinged door.

Use the instructions COMPLY FORCE, COMPLY TORQUE, STOP FORCE, and MOVE introduced in Section 9.1. In both cases draw a diagram showing the robot gripper, door, and $x, y, z$ co-ordinates.

9.7 For the following binary image calculate the center of area and axis of minimum moment of inertia:

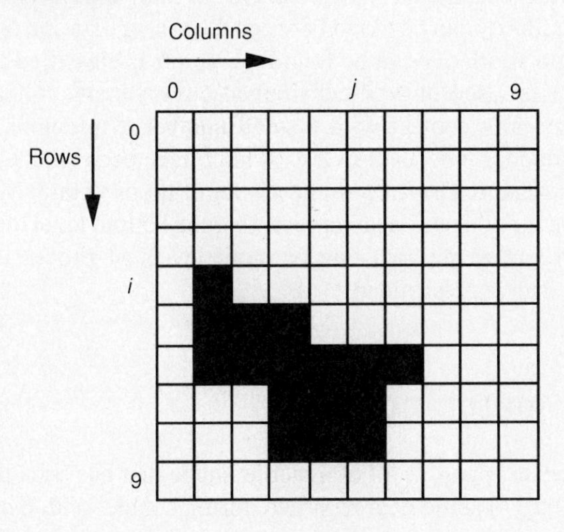

# 10 Pattern recognition of tactile data

Recognition of unknown objects is an essential first step in many grasping, manipulation, inspection, and assembly operations. This chapter describes some of the methods which have been used to classify tactile sensor data gathered from unknown objects by comparing them with prerecorded data from a number of known objects. An object is 'recognized' if a close similarity can be found between the sensor data and one of the sets of prerecorded data. If no similarity can be found the object is classified as unknown. These methods can only be used in well-constrained environments where the touch sensory system will come into contact with a small number of previously identified objects. Most of the techniques presented in this chapter have been adapted from work in the area of computer vision. They represent a bottom-up or data-driven approach where sensor information is collected (data representing the bottom level of the processing hierarchy) and then passed through one or more levels of processing until the 'meaning' of the sensor data is determined.

## 10.1 Template matching

For our purposes, a template is part or all of a tactile image that has been derived from contact with a known object. Tactile data recorded during contact with a new object is compared with one or more templates.

If the new sensor reading matches one of the templates, to some specified accuracy, then the object being investigated is assumed to be the same as the object corresponding to that template. In anything other than a highly structured environment, it would be unlikely that an object would have exactly the same position and orientation as it had during recording of the template. To allow for misalignment, a shift and compare procedure must be undertaken to find the position and orientation which minimizes the difference between the template and new sensor data. A tactile image would be represented as a two-dimensional array of data points. A simplified one-dimensional example of the matching process is illustrated in Figure 10.1. In this example the template $t$ is a sinusoid sampled at $i$ intervals. Measured data $m$ consists of $j$ samples. The measured data is compared with the template as the template is incrementally shifted right by an amount $x$. The difference $d_x$ between measured data and template shifted by $x$ may be calculated by summing absolute values of the difference between data points:

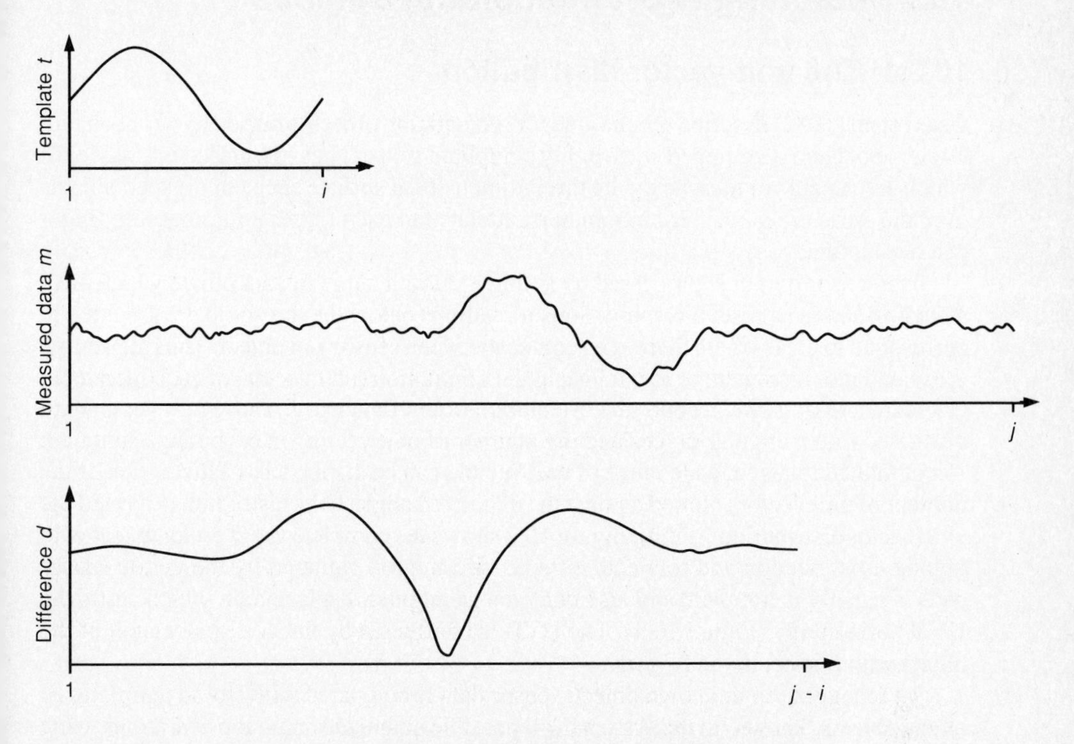


**Figure 10.1** An example — template matching of one-dimensional data

$$d_x = \sum_{y=1}^{i} \left| m_{(y+x)} - t_y \right|$$

Alternatively, the square of the differences between data points may be summed:

$$d_x = \sum_{y=1}^{i} (m_{(y+x)} - t_y)^2$$

The minimum value of difference $d_x$ shows the position of best match between measured data and template. When this technique is applied to a two-dimensional image, each template must be tested with incremental shifts in two orthogonal directions as well as incremental rotation at each point. This presents a large computational burden even for relatively low-resolution tactile data. Any method which reduces the amount of searching will greatly speed up the matching process. The following two matching techniques use two-dimensional silhouette images and, by transforming the data, template matching is reduced to a one-dimensional search.

## 10.2 Matching two-dimensional outlines

### 10.2.1 The unit-vector distribution

Ozaki et al. (1982) describe a technique for recognizing objects grasped by a robot hand. Their robot hand is equipped with highly compliant sensor pads (described in Chapter 6) which are capable of measuring the three-dimensional surface shape of grasped objects over the areas of contact. Potentiometers, attached to each finger joint, measure finger pad orientation.

The sensor output is processed by fitting a smooth curve to data points which have a reading above a preset threshold. Sensor readings below the threshold are assumed to correspond to areas where there is no contact between sensor and object. The fitted curve is divided into increments $s$ mm long and the orientation (unit vector) of each increment calculated, taking into account the orientation of the finger pad. These unit vectors are quantized into $n$ angular ranges and the number of unit vectors in each range summed. Accumulated totals for each range of unit vectors can be displayed as a histogram of the number of unit vectors plotted against the quantized angle. This histogram is termed the unit-vector distribution (UVD). Figure 10.2 shows the complete UVD for an object with oblong cross-section and rounded corners. Information gathered by the tactile sensor pads is usually incomplete and also contains an angular displacement which shifts the UVD horizontally (Figure 10.3). The UVD is unaffected by linear displacement of the object with respect to the fingers.

To recognize an unknown object, sensor data is compared with stored templates of all the objects 'known' to the system in all possible quantized angular orientations using a similar technique to that described in Section 10.1. A blurring operation on the stored

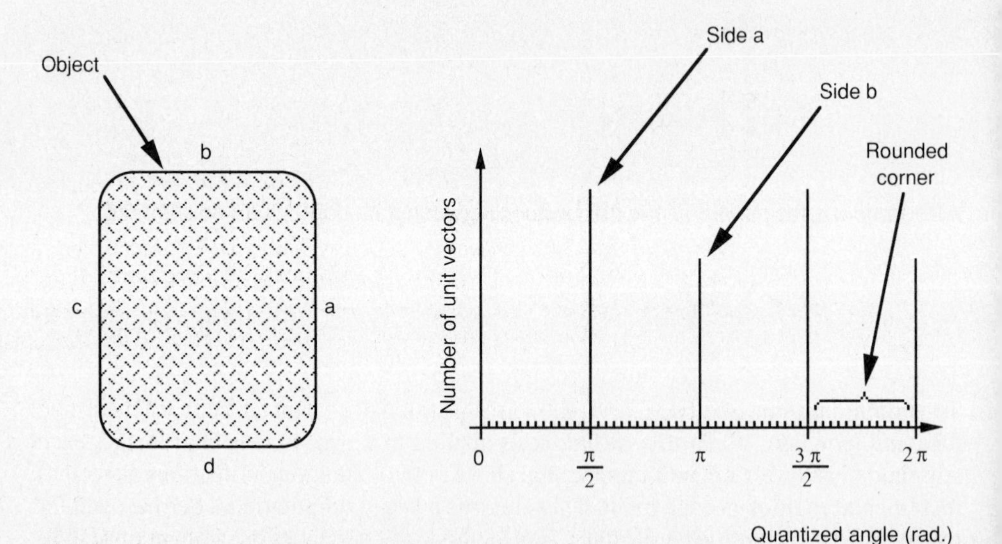


**Figure 10.2** The complete unit vector distribution for an object which is used as a template (Redrawn from Ozaki et al. 1982, ©1982 IEEE)

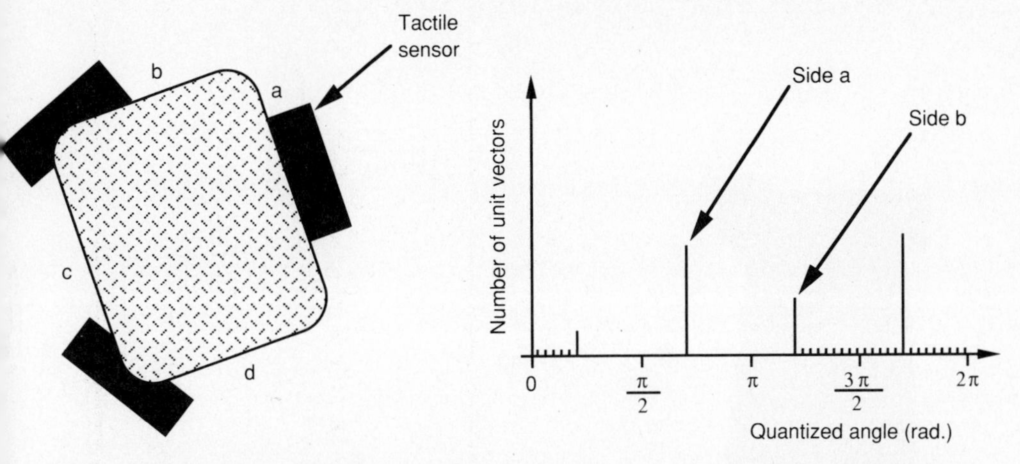


**Figure 10.3** Tactile sensor pads are used to measure part of the unit-vector distribution of an unknown object
(Redrawn from Ozaki et al. 1982, ©1982 IEEE)

template makes the matching operation tolerant of errors in the sampled data. Using this matching technique the system was able to discriminate cylinders and prisms.

For convex outlines the corresponding UVD is unique, but in the case of concave outlines the order in which the unit vectors occur is also required for identification, and this information is lost in the UVD. As described, the technique is applied to two-dimensional slices through an object. It may be possible to extend the technique to three-dimensional surface contours by dividing the surface into small planar surface patches.

## 10.2.2 Matching in $s - \theta$ space

This is another method of recognizing a two-dimensional silhouette image. In the original application silhouette data was gathered using the articulated whisker probe described in Chapter 7 (Russell 1985). Once again the outline is divided into short segments of equal length $s$ mm. By retaining the order of the unit vectors the previous restriction on concave objects is removed. Each segment is plotted in turn on a graph of distance around the outline $s$ against the angle of the outline tangent $\theta$. Figure 10.4 shows the silhouette image of a chess rook and the corresponding $s - \theta$ graph.

The $s - \theta$ representation has the properties that the shape of the curve is independent of the orientation of the object with respect to the sensor and both the shape and position of the curve are independent of linear displacement of the object with respect to the sensor. As the angle of orientation changes the curve is displaced vertically. Templates are taught to the system by recording the scanned outline of each object to be 'known' by the system. The $s - \theta$ information for an unknown object is identified by taking a segment of each template and comparing it with the measured data to find the best correspondence. To allow for a difference in orientation between object and template the mean value of the template and that of the sensor data are equalized before comparison. So:

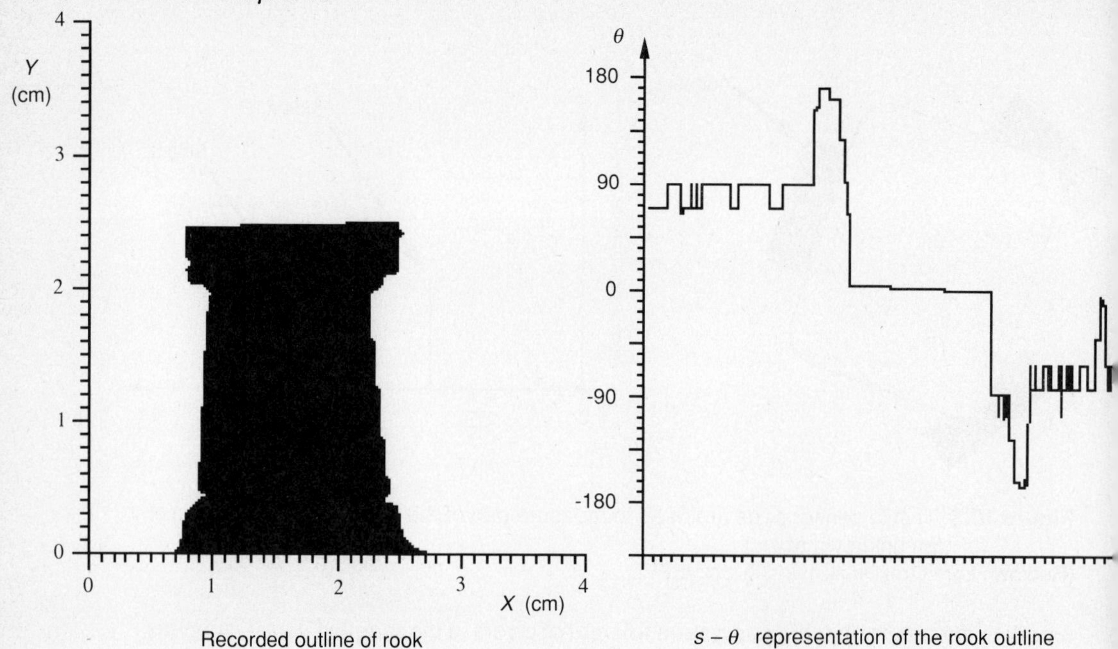


**Figure 10.4** The outline and $s - \theta$ representation of a chess rook

$$d_x = \sum_{y=1}^{i} \left| (m_{(y+x)} - m_{mean}) - (t_y - t_{mean}) \right| \tag{10.3}$$

where:

$$t_{mean} = \frac{\sum_{y=1}^{i} t_y}{i} \tag{10.4}$$

and:

$$m_{mean} = \frac{\sum_{y=1}^{i} m_{(y+x)}}{i} \tag{10.5}$$

When a successful match is achieved, the difference between the mean value of the template and mean value of the measured data indicates the angular displacement between template and measured object.

## 10.3 Object recognition using sparse data

A technique for object recognition using a small number of data points has been proposed by Grimson and Lozano-Pérez (1984). The technique requires an accurate geometric model of all known objects and assumes a tactile sensor with some form of mobility. This sensor must be able to locate contact points accurately and the surface normal at the point of contact with somewhat lower accuracy. The method only works for polygons but can be applied in three dimensions, although this description will be limited to the two-dimensional case.

Consider a planar object $O_j$ which has $n_j = 4$ sides. When the tactile sensor makes its first contact at point 1 (P1) that contact could be on any of the four sides. Similarly, the second contact is made at point 2 (P2) and that too could be on any of the four sides — thus giving sixteen possibilities. This is shown diagrammatically in the interpretation tree (IT) (Figure 10.5). The IT shows all possible pairings of contact points with object faces. A path from top to bottom of the tree represents one possible interpretation of the sensor data. There is a different tree for each object 'known' to the system. For $s$ points on $n$ faces there will be $n^s$ possible interpretations. The idea of this technique is to use local constraints to remove invalid pairings of contact points with object faces and thus simplify the ITs:

1. *Distance* — the distance between each pair of contact points Pi must be a possible distance between the faces paired with them in an interpretation.

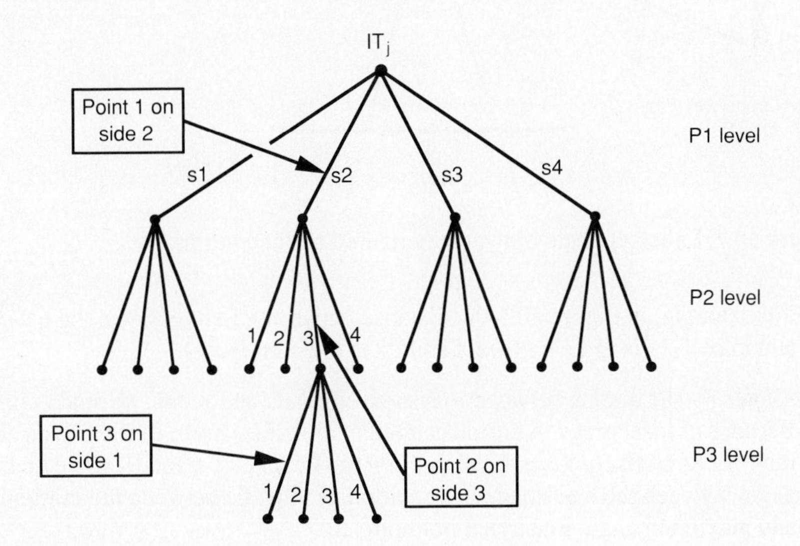


**Figure 10.5** The interpretation tree
(Redrawn from Grimson and Lozano-Pérez 1984, ©1984 MIT Press)

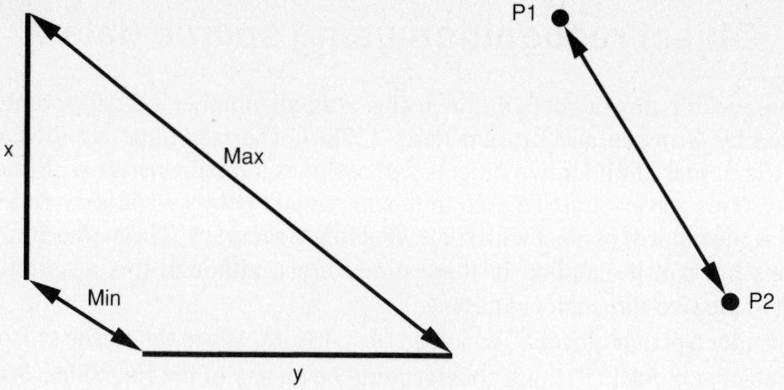

**Figure 10.6** Feasible distance between contact points

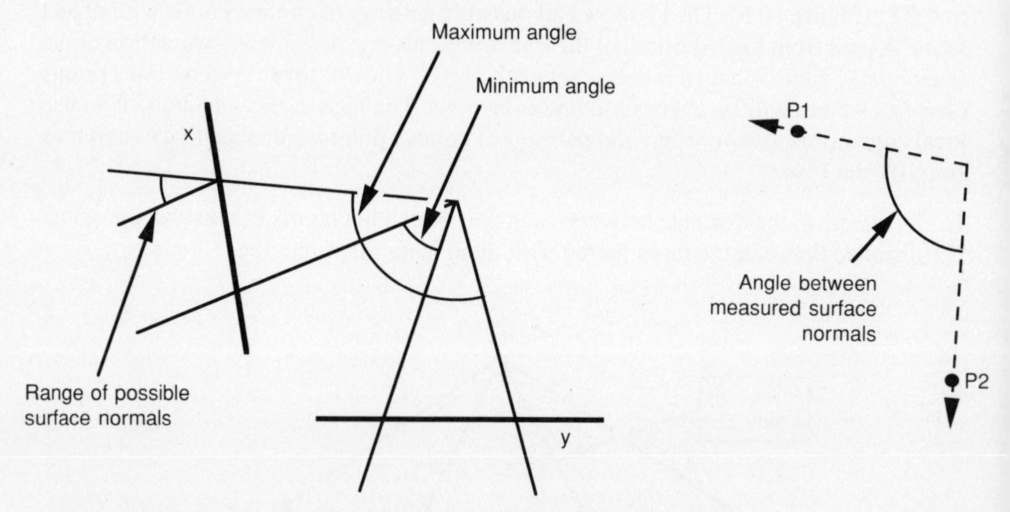


**Figure 10.7** Feasible angle between measured surface normals

For example, in Figure 10.6 the distance between P1 and P2 must be between min and max if P1 is to lie on face x and P2 is to lie on face y.

2. *Angles* — the angles between measured normals and model normals must agree. Because of inaccuracy in the measurement of surface normals there is an allowable range of normals for faces x and y. For P1 to lie on face x and P2 to lie on face y the angle between the measured surface normals must lie between the minimum angle and maximum angle illustrated in Figure 10.7.

3. *Orientation* — the orientation between the surface normal at a contact point and a vector between that contact point and another must lie within the feasible range (Figure 10.8).

All pairings of contact points with faces which fail one or more of these tests are pruned from the IT. After a number of touch points have been recorded and the IT pruned to a

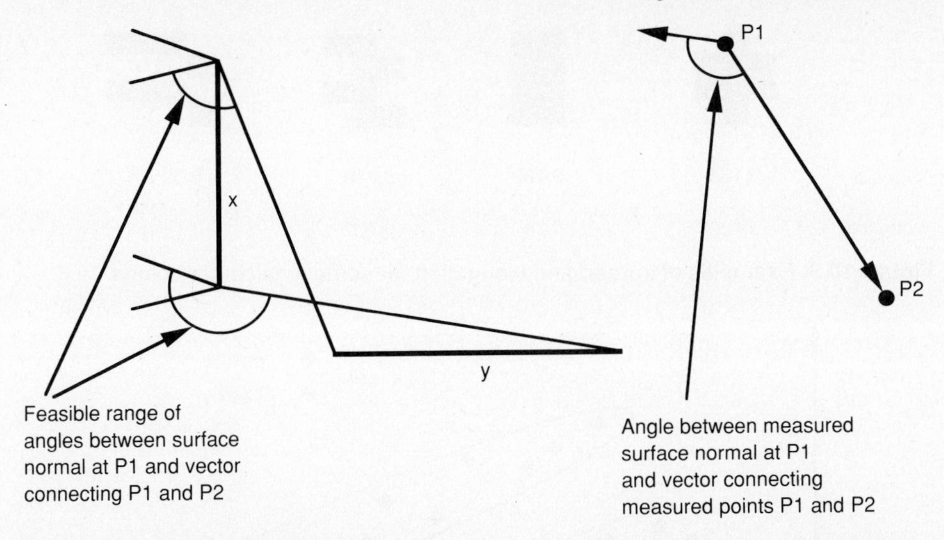


**Figure 10.8** Feasible angle between the surface normal at point P1 and a vector between points P1 and P2

small number of interpretations, the remaining interpretations are tested by determining a rotation and translation of the template which would match the recorded data. If there is more than one possible interpretation then more data must be gathered.

## 10.4 Classification of binary tactile images

We have seen in Chapter 9 that a binary image derived from a tactile sensor array can be used to determine the position and orientation of a gripped object. Different objects produce different binary images and may by distinguished by comparing features derived from the images (Winston and Horn 1984). Features for classifying images should be independent of the orientation of the image on the sensor. For example, object area is a feature which is independent of rotation. Spread and elongation are two additional quantities, derived from moments of area (see Chapter 9), which are also independent of rotation:

$$\text{Spread} = \frac{(a+c)}{m_{00}^{2}} \tag{10.6}$$

$$\text{Elongation} = \frac{\sqrt{b^{2}+(a-c)^{2}}}{(a+c)} \tag{10.7}$$

where:

$$a = m_{20} - m_{10}^{2}/m_{00}^{2} \tag{10.8}$$

$$b = m_{11} - (m_{10} \cdot m_{01})/m_{00}^{2} \tag{10.9}$$

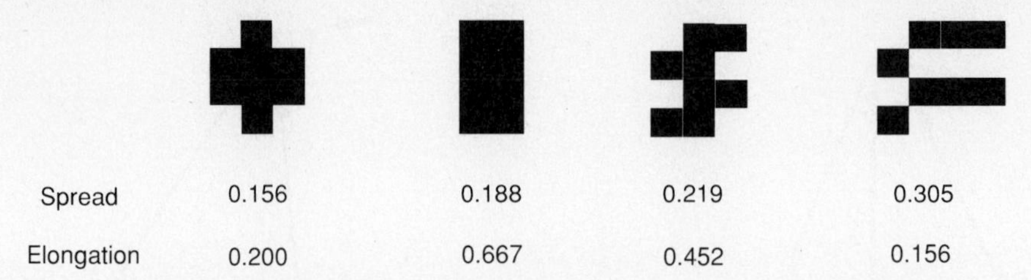


**Figure 10.9** Examples of spread and elongation for some small binary blobs

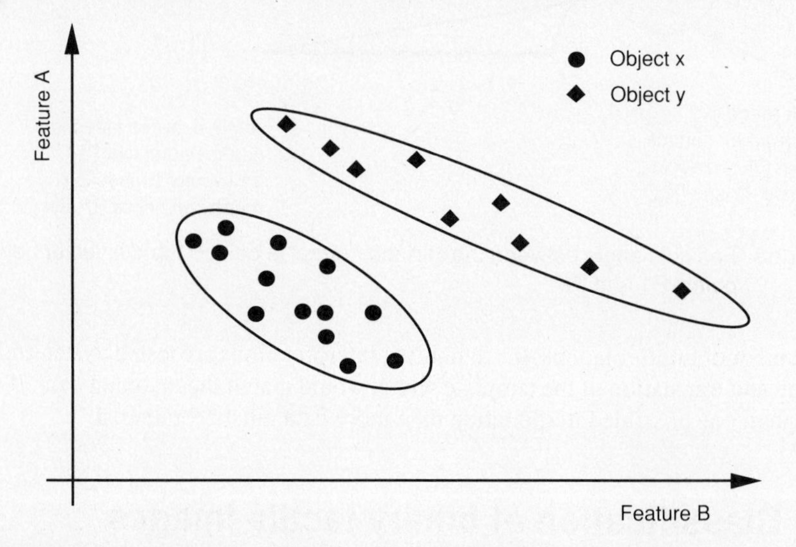


**Figure 10.10** The grouping of objects in feature space

$$c = m_{02} - m_{01}^2 / m_{00}^2 \tag{10.10}$$

As the name suggests, spread is a measure of image compactness and elongation indicates whether the image is spread along a preferred axis. Figure 10.9 shows examples of small binary images and their corresponding values of spread and elongation. Additional features such as perimeter may also be required to improve discrimination. For a particular application, *n* features may be selected for the classification process. These features can be thought of as defining an *n*-dimensional space with measured values of the features representing co-ordinates of points in that space.

Because of quantization errors and measurement noise there will be variation in the measured value of each feature — for the same object. If suitable features have been chosen then points corresponding to each object will cluster together as illustrated in Figure 10.10. A simple way of classifying an unknown object is to calculate the Euclidean distance between the point in feature space for that object and the position for each of the known objects. The unknown object will be assumed to be the same as the nearest known object. Many other techniques are available to perform classification (Duda and Hart 1973).

Similar pattern classification techniques have been applied to data obtained by grasping an object with a robot gripper (Kinoshita et al. 1975). Finger joint angles and the pattern of contact on an array of binary sensors were used as co-ordinates for a multi-dimensional space. Unfortunately, it appears that only a restricted range of shapes can be discriminated under these circumstances.

## 10.5 Conclusion

With vision sensors it is usually possible to get a complete view of one side of an object in a single image either by choosing a suitable viewpoint or adjusting the lens field of view. During visual image processing the flow of information tends to be mainly upwards, from the raw data to higher and more abstract forms of representation — the ultimate goal is a description of the position, orientation and identity of the object. Tactile sensors can only modify their field of view by actively exploring objects which are larger than the sensor area. One approach is to simulate a larger sensor by prodding the unknown object in a raster scan fashion and thus build up a suitably sized image. A more interesting scheme is to use the sensor to actively search for tactile information by tracing significant features of the object. This active sensing technique is considered in the next chapter.

## Bibliography

Duda, R. O., and Hart, P. E., *Pattern Classification and Scene Analysis*, John Wiley & Sons, New York, 1973.

Grimson, W. E. L., and Lozano-Pérez, T., 'Model-Based Recognition and Localization From Sparse Range or Tactile Data', *The International Journal of Robotics Research*, Vol. 3, No. 3, Fall 1984, pp. 3–35.

Kinoshita, G.-I., et al., 'Pattern Classification by Dynamic Tactile Sense Information Processing', *Pattern Recognition*, Pergamon Press, Vol. 7, 1975, pp. 243–51.

Ozaki, H., et al., 'Pattern Recognition of a Grasped Object by Unit-Vector Distribution', *IEEE Transactions on Systems, Man and Cybernetics*, Vol. SMC-12, No. 3, May/June 1982, pp. 315–24.

Russell, R. A., 'Object Recognition Using Articulated Whisker Probes', *Proceedings of the 15th ISIR*, Tokyo, Japan, 1985, pp. 605–12.

Winston, D. E., and Horn, B. K. P., *Lisp*, (2nd ed.), Addison-Wesley, Reading, Massachusetts, 1984.

## Questions

10.1 For the following one-dimensional data find the shifted position of the template which gives the best match between data and template.

Template (25 values)

| | | | | | | | | | |
|---|---|---|---|---|---|---|---|---|---|
| 11 | 9 | 8 | 9 | 11 | 13 | 11 | 8 | 6 | 7 |
| 14 | 23 | 29 | 29 | 23 | 14 | 7 | 6 | 8 | 11 |
| 13 | 11 | 9 | 8 | 9 | | | | | |

Data (146 values)

| | | | | | | | | | |
|---|---|---|---|---|---|---|---|---|---|
| 10 | 9 | 9 | 8 | 7 | 9 | 11 | 12 | 11 | 12 |
| 11 | 9 | 8 | 9 | 8 | 10 | 9 | 9 | 10 | 11 |
| 10 | 10 | 8 | 10 | 9 | 7 | 10 | 11 | 9 | 10 |
| 11 | 12 | 12 | 10 | 11 | 12 | 13 | 14 | 12 | 10 |
| 12 | 11 | 10 | 12 | 11 | 11 | 11 | 10 | 8 | 10 |
| 12 | 13 | 12 | 12 | 12 | 11 | 10 | 10 | 10 | 12 |
| 13 | 13 | 14 | 12 | 13 | 14 | 14 | 14 | 12 | 12 |
| 13 | 13 | 11 | 12 | 13 | 15 | 14 | 11 | 12 | 12 |
| 13 | 10 | 9 | 5 | 6 | 12 | 23 | 31 | 29 | 23 |
| 15 | 8 | 9 | 11 | 14 | 16 | 12 | 11 | 9 | 10 |
| 12 | 12 | 12 | 10 | 11 | 12 | 12 | 11 | 12 | 12 |
| 10 | 9 | 11 | 10 | 9 | 8 | 9 | 8 | 10 | 11 |
| 10 | 10 | 8 | 7 | 7 | 6 | 6 | 8 | 9 | 9 |
| 9 | 9 | 8 | 7 | 7 | 7 | 6 | 5 | 5 | 8 |
| 8 | 8 | 8 | 7 | 6 | 8 | | | | |

10.2 Describe those characteristics of the unit-vector distribution (UVD) which make it suitable for use in template matching.

10.3 Sketch the UVD for an object having a circular outline.

10.4 Describe those characteristics of the $s-\theta$ representation which make it suitable for use in template matching.

10.5 Sketch the $s-\theta$ representation for an object having a circular outline.

10.6 Consider the following planar object:

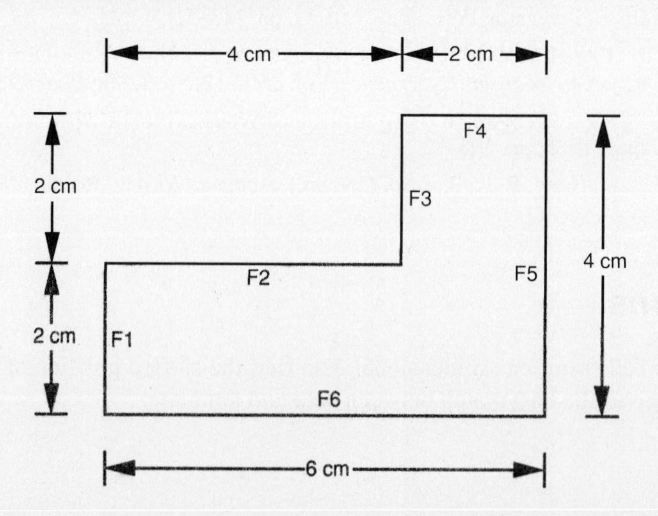

Complete the following table which contains the minimum and maximum separations for two points (point A and point B) lying on every possible pairing of object faces (faces F1 to F6). (*Hint*: use a scale drawing and rule to work out the separations.)

| | | Point A | | | | | |
|---|---|---|---|---|---|---|---|
| | | F1 | F2 | F3 | F4 | F5 | F6 |
| | F1 | (0,2) | ? | ? | ? | ? | ? |
| | F2 | ? | (0,4) | (0,4.5) | ? | ? | ? |
| Point B | F3 | ? | (0,4.5) | ? | ? | (2,4.4) | ? |
| | F4 | ? | ? | ? | ? | ? | ? |
| | F5 | ? | ? | ? | (2,4.4) | ? | ? |
| | F6 | ? | ? | ? | ? | ? | ? |

On your table indicate all possible pairings of points with faces for points P1 and P2 separated by 7 cm.

10.7 During a training session features A, B, and C have been recorded from tactile images of a number of *gadgets*, *objects* and *widgets*. The results are as follows:

| Gadget | A | B | C |
|---|---|---|---|
| 1 | 14.9 | 7.7 | 46.1 |
| 2 | 13.8 | 8.6 | 46.2 |
| 3 | 13.2 | 9.0 | 44.3 |
| 4 | 16.0 | 7.2 | 45.1 |
| 5 | 16.8 | 8.3 | 44.7 |
| 6 | 13.2 | 8.7 | 45.2 |
| 7 | 14.5 | 6.9 | 44.4 |

| Object | A | B | C |
|---|---|---|---|
| 1 | 14.4 | 4.6 | 46.2 |
| 2 | 12.0 | 7.4 | 49.0 |
| 3 | 12.5 | 7.0 | 48.2 |
| 4 | 13.1 | 4.6 | 49.8 |
| 5 | 12.4 | 5.7 | 50.2 |
| 6 | 13.3 | 5.8 | 48.6 |
| 7 | 12.8 | 6.3 | 48.9 |

| Widget | A | B | C |
|---|---|---|---|
| 1 | 15.6 | 8.8 | 49.6 |
| 2 | 17.6 | 7.0 | 49.2 |
| 3 | 16.8 | 7.2 | 47.4 |
| 4 | 16.0 | 7.3 | 49.1 |
| 5 | 17.3 | 5.9 | 48.3 |
| 6 | 16.3 | 6.3 | 48.6 |
| 7 | 16.4 | 8.3 | 48.0 |

Features A, B, and C were then recorded for four unknown items and the results were as follows:

| Item | A | B | C |
|---|---|---|---|
| w | 11.9 | 8.4 | 47.2 |
| x | 14.4 | 9.0 | 46.2 |
| y | 15.7 | 5.3 | 46.3 |
| z | 16.3 | 5.3 | 50.1 |

Determine the identity of items w, x, y, and z.

# 11 Active touch sensing

Chapter 10 described methods of using tactile information for object recognition. Most of these techniques are based on algorithms developed for processing visual images. The field of view of a vision system can usually be adjusted to give a complete image of an object. The sensing process therefore involves taking a picture of the unknown object and then processing the resulting data to identify the object. This sensing paradigm does not match the characteristics of tactile sensing. The size of a tactile sensor array will usually be much smaller than the object being examined. Therefore, a full image of the object cannot be formed from the results of one contact with the object. By contrast with vision, touch sensing requires active exploration of an object to trace out surface contours and test mechanical properties of the constituent materials.

Active touch sensing takes place on two levels. At a low level, many object properties can be derived by simple tests (like squeezing a pear to see if it is ripe, or measuring the size of a plate by tracing its outer edge with a finger). At a higher level, the top-down sensing approach uses a repeated hypothesis verification loop to identify unknown objects. This strategy seems to match the characteristics of tactile sensors. Hypotheses relating to possible interpretations of the data are generated and then tested to examine their consistency. An interpretation is sought which accounts for all of the observed data and provides a consistent explanation for the sensor data. If a consistent interpretation cannot be found then more sensing movements are planned to gather information which can help to resolve the ambiguity and provide a unique interpretation for all the sensor data.

## 11.1 Measurement of tactile properties

A wide range of material properties can be measured by tactile sensing. They can be categorized depending upon the amount of 'intelligence' required to supervise the measurement procedure. At the very lowest level, simple predefined sequences of actions are used to determine *tactile primitives*. The term 'tactile primitive' and the terms 'tactile feature' and 'exploratory procedure' which are introduced later in this chapter were coined by workers at the University of Pennsylvania (Stansfield 1986). Tactile primitives are properties of objects which can be directly inferred from sensor data. Table 11.1 shows a range of tactile primitives and the sensor data necessary to measure them.

In some cases, such as object temperature or mass, the thermistor or load cell output

**Table 11.1** Tactile properties which can be directly inferred from sensor data

| | Tactile primitives | | | | |
|---|---|---|---|---|---|
| Groupings | Mechanical behavior | Gross characteristics | Surface shape | Thermal | Electrical behavior |
| Examples | Compliance<br>Resilience<br>Viscosity | Size<br>Mass | Corner<br>Edge<br>Surface normal | Temperature<br>Thermal conductivity<br>Thermal diffusivity | Electrical conductivity<br>Contact potential |
| **Parameters to be measured** | | | | | |
| Forces and torques transmitted between gripper and object | ✓ | ✓ | | | |
| Image of area of contact between gripper and object | | | ✓ | | |
| Configuration of robot gripper | ✓ | ✓ | ✓ | | |
| Thermal | | | | ✓ | |
| Electrical | | | | | ✓ |

directly represents the required quantity. Other tactile primitives are determined by performing simple measurement procedures (see Figures 11.1, 11.2, and 11.3):

### 11.1.1 Estimation of compliance

Compliance is a measure of the amount by which a material, or structure, yields when a force is applied. To relate this measured quantity to our own experience the applied force $F$ would be limited to the non-destructive magnitude we would apply when testing the hardness of an unknown object. Any object will yield a significant amount if sufficient force is applied!

The method involves measuring the increased depth of indentation caused by an increase in applied force. This is illustrated in Figure 11.1 and a step-by-step procedure for performing the measurement is given below:

Procedure for estimating COMPLIANCE

Move until a force $F$ is applied to the object.
Store present indentation depth as $D1$.
Press harder until a force of $2 \cdot F$ is applied to object.
Store present indentation depth as $D2$.

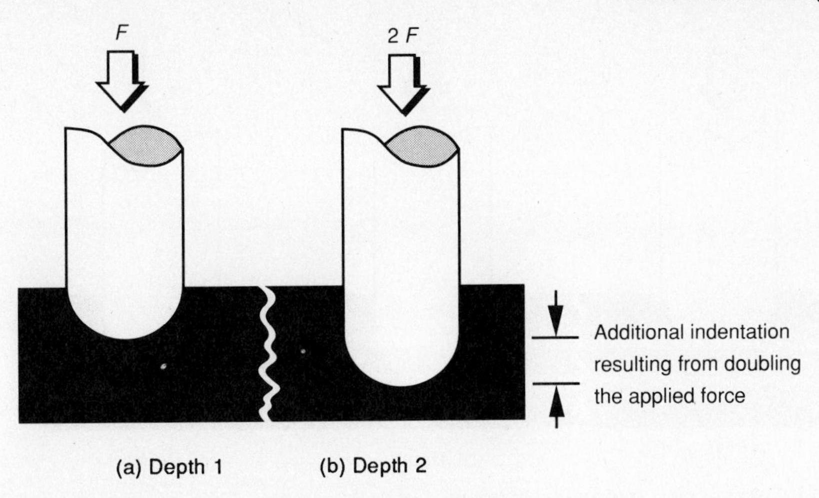


**Figure 11.1** Estimation of material compliance

The distance between *D1* and *D2* is a measure of compliance.

End

To make a valid measurement the compliance of the robot, associated probe, and the support for the material sample must all be significantly lower than the compliance of the material being tested.

### 11.1.2 Estimation of resilience

Resilience indicates how well a material returns to its original shape after being deformed.

This quantity can be assessed by increasing applied force and then reducing to the original value (Figure 11.2). How closely the probe assumes its original depth is a measure of resilience. The following procedure for measuring resilience is an extension of the procedure for assessing compliance.

Procedure for estimating RESILIENCE

Move until a force *F* is applied to the object.
Store present indentation depth as *D1*.
Press harder until a force of 2·*F* is applied to object.
Store present indentation depth as *D2*.
Move away from surface until applied force falls to *F*.
Store present indentation depth as *D3*.
$(D2 - D3)/(D2 - D1)$ is a measure of material resilience.

End

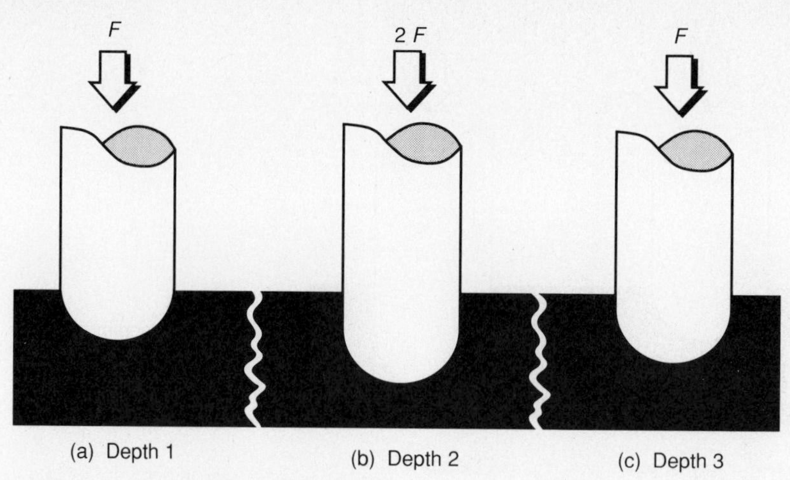


**Figure 11.2** Assessing material resilience

## 11.1.3 Estimation of pliability

Some materials deform slowly over time when subjected to a continuing force. This viscoelastic behavior is reversible in cross-linked polymer materials such as natural and synthetic rubbers. When the deformation is irreversible the effect is termed viscosity. This behavior is seen in liquids such as treacle and pitch. In metals the slow, irreversible deformation over time is called creep. Because a number of different terms are used to describe a similar effect, I will use the word *pliability* to indicate deformation of a material over time resulting from a constant force as measured by the procedure described below (illustrated in Figure 11.3).

During lifting and carrying maneuvers it is essential to monitor gripping forces. Objects with a relatively high pliability can deform due to applied forces. If deformation of the object is not accommodated by tightening the grip, gripping forces could fall to a point where the object starts to slip. Pliability can be assessed by the following procedure:

Procedure for estimating PLIABILITY

> Move until a force $F$ is applied to the object.
> Store present indentation depth as $D1$.
> Wait for a fixed time.
> Store present indentation depth as $D2$.
> $(D2 - D1)$ is a measure of pliability.

End

## 11.1.4 Thermal characteristics

In Chapter 7 a thermal sensor was described based on the human thermal sense. The sensor output can be used to calculate $D$, the percentage temperature drop towards

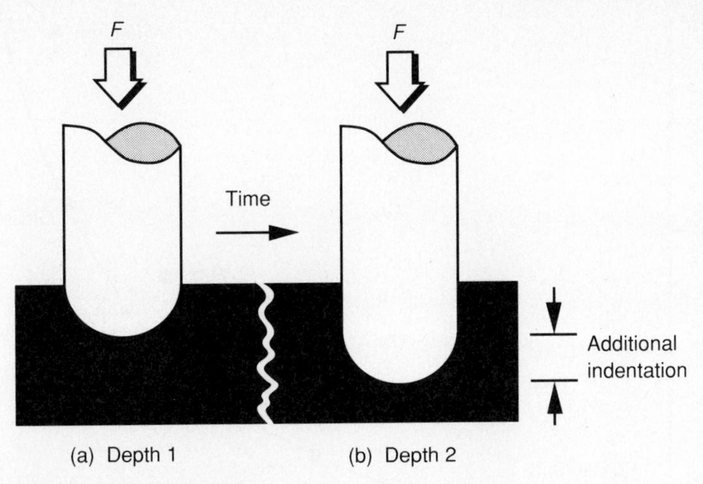


**Figure 11.3** Gauging material pliability

ambient temperature a short time after contact with a material (see Section 7.2). This temperature drop in the thermal sensor, when the sensor touches an external object, depends upon the thermal conductivity and thermal diffusivity of the material forming the object (Russell and Paoloni 1985). An unknown material can be categorized in terms of a range of prerecorded materials if the value of $D$ measured after contact with the unknown material is compared with values of $D$ recorded for the known materials.

A method of directly inferring other material properties from measured values of $D$ has been suggested by examining a graph of $D$ plotted against material specific mass (Figure 11.4). The plotted points fall close to a smooth curve. Therefore, using the information in this graph, an estimate of the material specific mass can be deduced from the percentage temperature drop towards the ambient temperature.

Many tactile primitives provide information about the materials making up an object. Each primitive, on its own, may not provide sufficient information to positively identify a particular material. However, the combination of a number of tactile primitives will improve the accuracy of identification. Figure 11.5 shows that a range of different materials can be discriminated based on measurements of compliance and resilience.

## 11.2 Tactile features

*Tactile features* are at a higher level in the hierarchy of data representations. These are also properties of the object, but they require some level of intelligent, guided search to gather the necessary data. Table 11.2 lists some tactile features.

Tactile features may consist of an extended surface region or the relative motion between parts of an object. In both cases, an intelligent, guided search will be required to discover the nature and extent of the feature. This kind of sensing scheme, involving a sensor-guided search, has been called an *exploratory procedure* (Stansfield 1986).

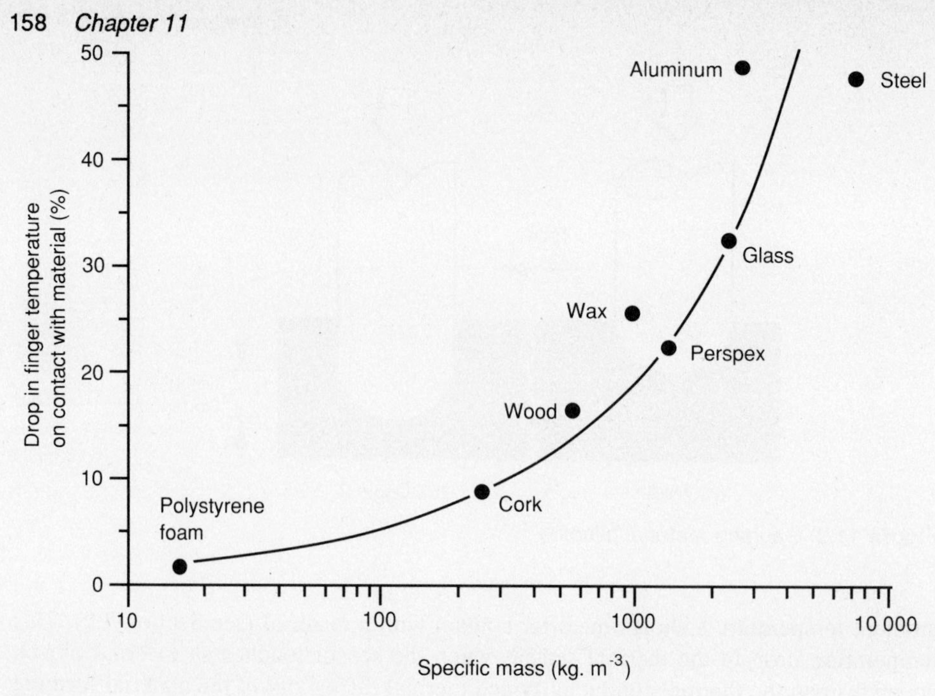


**Figure 11.4** Material specific mass plotted against sensor temperature drop towards ambient temperature three seconds after contact with the material
(Redrawn from Russell 1988, courtesy Cambridge University Press)

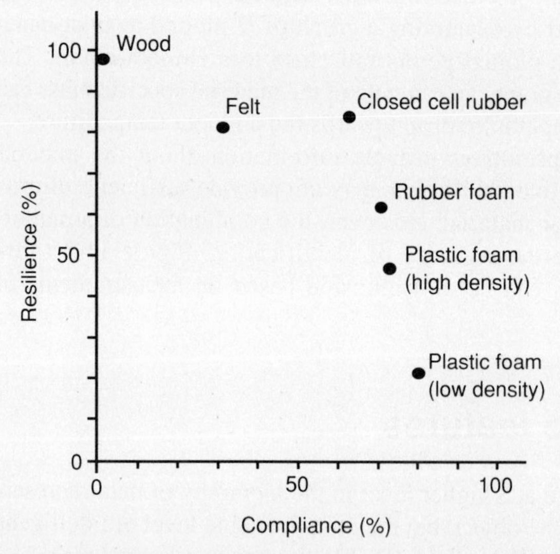


**Figure 11.5** A range of common materials can be discriminated by comparing values of compliance and resilience measured using the algorithms described in this Chapter.

**Table 11.2** Some tactile features of objects

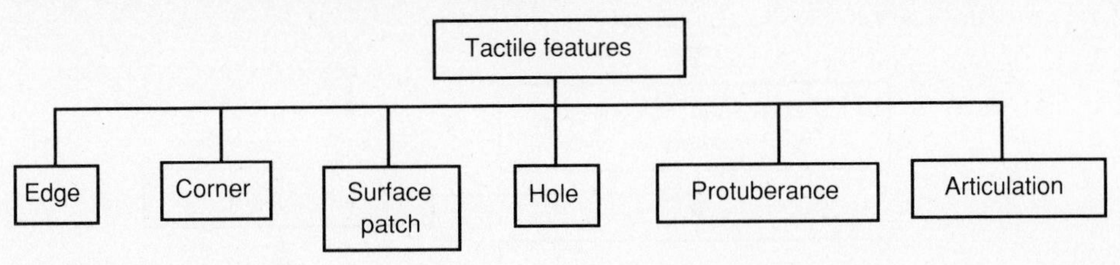


The flowchart shown in Figure 11.6 details a simple exploratory procedure to find and trace the two-dimensional outline of an object using a mobile tactile sensor probe. This procedure starts with a raster scan which is used to search a wide area. The separation between successive scans is made smaller than the minimum dimension of the objects being sought and the raster scan ends when the search area has been covered or initial contact is made with an object.

If contact has been made with an object then control of the sensor movement changes to edge-tracking mode and the object outline is traced in an anticlockwise direction. Following each contact the sensor is withdrawn a short distance and then moved parallel to the object outline. If contact is made during this parallel movement, the outline is assumed to be curving to the right. Otherwise, the sensor is moved left to confirm the location of the outline. If contact is not made during this movement, the outline is assumed to be curving to the left. The search continues until the sensor arrives back at the original point of contact, in which case the complete outline of the object has been traced, or until contact with the object is lost. Figure 11.7 illustrates the raster scan and boundary tracing trajectories.

There is evidence that tactile features are very important in human object recognition. Figure 11.8 shows recorded human index finger trajectories while exploring a range of simple shapes (Gurfinkel et al. 1974). The finger spends little time on smooth surfaces, concentrating on tracing edges and exploring the surface close to vertices. Therefore, active sensing is important at the level of gathering tactile sensory data. Active sensing is also important at the level of object recognition.

## 11.3 The active approach to object recognition

The methods of analyzing tactile information, presented in Chapter 10, follow a bottom-up or data-driven approach. In such a scheme, three steps are followed in sequence. Previous steps are not repeated once they have been completed. Sensor data is collected, processed to extract useful statistics, and then these results are categorized to establish the identity of the unknown object. This process is illustrated in Figure 11.9(a).

Rather than start by gathering every conceivable piece of information about the unknown object, an alternative is to start with a hypothesis of what the object might be (Hillis 1981). The assumed identity of the object will be that which provides the best fit

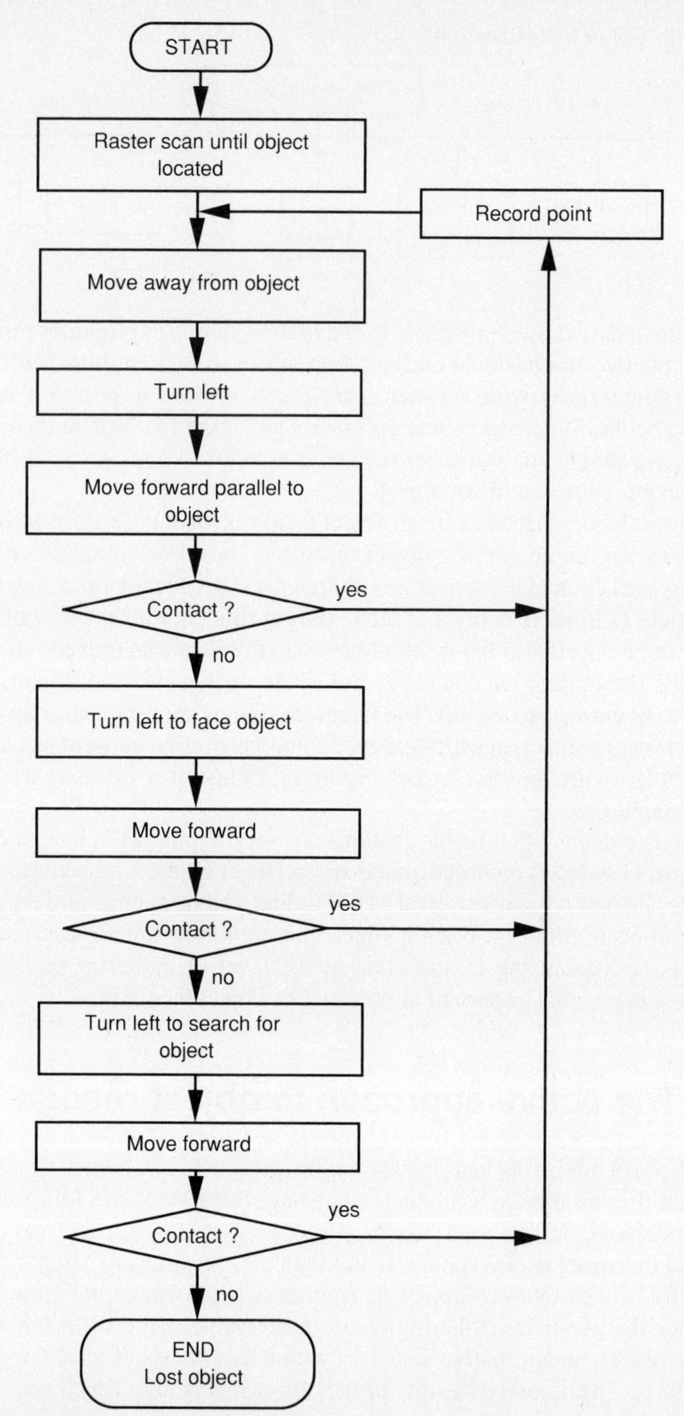


**Figure 11.6** The outline tracing strategy
(Redrawn from Russell 1985)

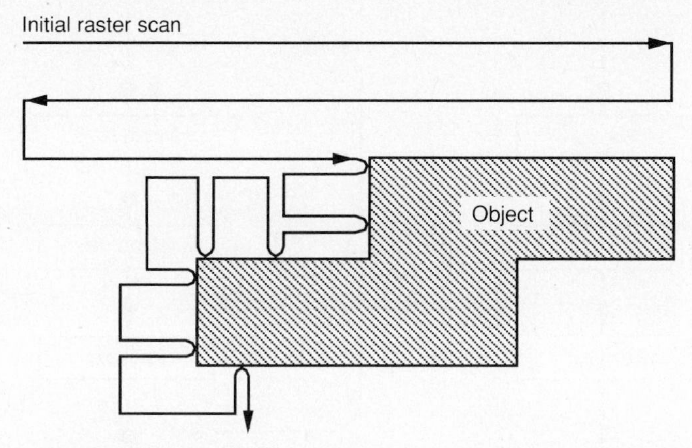


**Figure 11.7** Tracing a two-dimensional outline

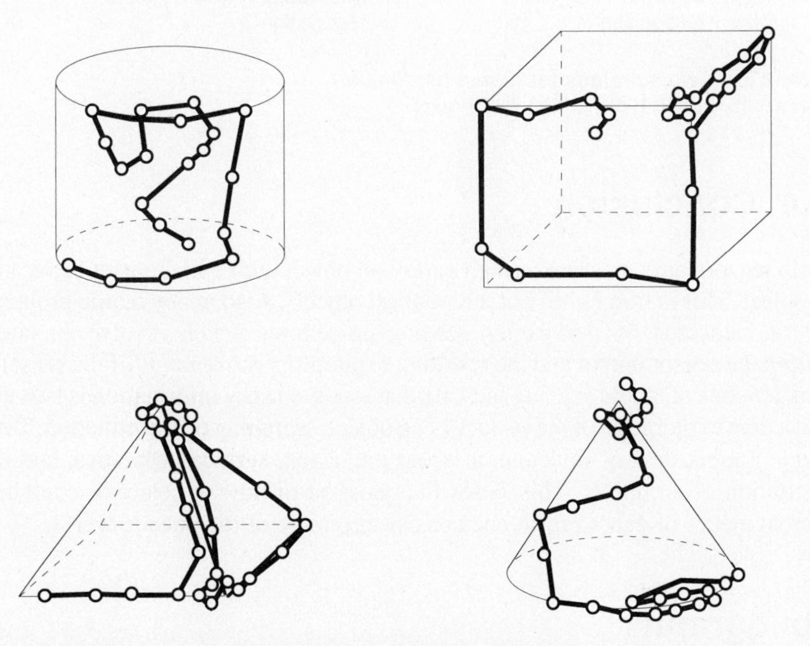

**Figure 11.8** Paths of finger movement during object exploration
(Redrawn from Gurfinkel et al. 1974, courtesy Scripta Technica, Inc.)

to the available data. The hypothesis is then tested by gathering additional information about the object. After analysis the additional data may support the original hypothesis or suggest a new one which provides a better fit to all of the available data. The hypothesize and test loop (illustrated in Figure 11.9(b)) is repeated until the present hypothesis and data agree. Data gathering is driven by the need for that particular piece of information. This approach is economical in the tactile sensing situation where information is gathered a little at a time.

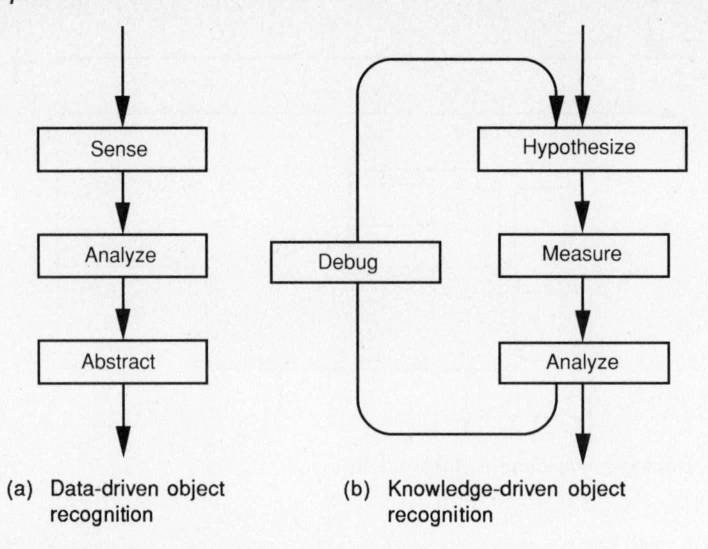


**Figure 11.9** Two schemes for object recognition
(Redrawn from Hillis 1981, ©1981 MIT Press)

# 11.4 Conclusion

Tactile sensors only measure a small part of an object with each contact between sensor and object. This is true for all but the smallest objects. Also, many tactile properties can only be measured by performing sensing procedures which involve an interaction between the sensor output and the resulting exploratory movements of the sensor. These characteristics of tactile sensing indicate that touch sensory information is best gathered by an active exploration of the surface of an object, searching out information. Touch can provide a great deal of information about the shape, surface properties, and material constitution of an object. This raises the question of how to select or combine these different pieces of data to form one consistent picture of the outside world.

# Bibliography

Bajcsy, R., et al., 'What Can We Learn From One Finger Experiments?', *Grasp Lab Report MS-CIS-83-03*, Department of Computer and Information Science, University of Pennsylvania, August 1983.

Dario, P., and Buttazzo, G., 'An Anthropomorphic Robot Finger for Investigating Artificial Tactile Perception', *The International Journal of Robotics Research*, Vol. 6, No. 3, Fall 1987, pp. 25–48.

Gurfinkel, V. S., et al., 'Tactile Sensitizing of Manipulators', *Engineering Cybernetics*, Vol. 12, Nos. 4–6, 1974, pp. 47–56.

Hillis, W. D., 'A High Resolution Imaging Touch Sensor', *The International Journal of Robotics Research*, Vol. 1, No. 2, Summer 1982, pp. 33–44.

RUSSELL, R. A., 'Object Recognition Using Articulated Whisker Probes', *Proceedings of the 15th International Symposium on Industrial Robots*, Japan, 1985, pp. 605–612.

RUSSELL, R. A., 'Thermal Sensor for Object Shape and Material Constitution', *Robotica*, Vol. 6, 1988, pp. 31–34.

RUSSELL, R. A., and PAOLONI, F. J., 'A Robot Sensor for Measuring Thermal Properties of Gripped Objects', *IEEE Transactions on Instrumentation and Measurement*, Vol. IM-34, No. 3, September 1985, pp. 458–460.

STANSFIELD, S. A., 'Primitives, Features and Exploratory Procedures: Building a Robot Tactile Perception System', *Proceedings of IEEE International Conference on Robotics and Automation*, San Francisco, 1986, pp. 1274–9.

## Questions

11.1 Explain the major differences in characteristics between computer vision and tactile sensing.

11.2 Describe a procedure for estimating the compliance of a material by indenting the material with a tactile probe.

11.3 Write and explain the operation of a procedure to assess the coefficient of friction between a tactile probe and external object.

11.4 Explain how thermal measurements may be used to indicate the specific mass of the material making up an object.

11.5 Explain the terms 'tactile primitive' and 'tactile feature' as they are used in this chapter.

11.6 Describe an exploratory procedure for tracing the two-dimensional outline of an object using a tactile probe.

11.7 A robot is to locate and switch on a light switch. Which tactile features of a light switch should be measured to assist:

(a) recognition; and
(b) manipulation of the switch?

Justify your selection.

11.8 Describe the hypothesize and test approach to object recognition.

# 12 Combining sensory data from multiple sources

Currently, it would be the exception rather than the rule to find an industrial robot system containing any form of advanced sensing. In the relatively few cases which do exist, the sensors are used in a very constrained manner. The number of different quantities to be measured is usually small, and specific sensors are chosen in advance and carefully positioned and calibrated in order to measure them. This is possible in a closely controlled environment and where the robot performs a small number of well-defined tasks.

Looking to the future, what will be the sensing requirements for autonomous, multi-function robots operating in ill-defined environments? It seems probable that they will be provided with a range of sensors including computer vision, infra-red imaging, laser range finding, etc., in addition to tactile sensing. Faced with the problem of acquiring a specific item of sensory data, the robot system will have to find out which sensor, or sensors, can provide the information. In the event that only one suitable sensor exists, then there is no option but to use that single source of information.

## 12.1 Multiple sensors — choose or fuse?

If several sensors are available, all of which can provide a measurement of the required data, the problem arises of how to use these multiple information sources to the best advantage. There seem to be two main approaches to this problem: either choose the single most appropriate sensor and rely on its output alone, or combine the data from all applicable sensors to generate a composite view (Harmon et al. 1986).

## 12.2 Choosing between sensors

Choosing between a number of sensors, perhaps guided by data from other sensors, is sometimes termed sensor integration. Here are some suggestions for sensor selection criteria. Such considerations could be used to choose the single, most appropriate, sensor for measuring a desired quantity:

- *Sensor reliability* — each sensor measures to a certain accuracy and is susceptible to drift, interference, and random effects. Under unfavorable circumstances sensors may give completely erroneous readings (ultrasonic sonar sensors cannot detect smooth flat surfaces when approached from an angle — the ultrasound bounces off the wall in such a way that it is not returned to the receiver).

- *Sensor operating limits* — sensors only register correctly over a limited range of the measured quantity (one commercially available ultrasonic sonar sensor can detect objects at distances between 25 cm and 1000 cm).
- *Operating cost* — there may be some kind of penalty associated with using a particular sensor. The sensor may consume a large amount of energy when operating, take a significant length of time to gather data, or require extensive data processing to interpret the resulting data. A battlefield robot may have to balance the possibility of detection when using an active sensor against the less useful information that it can derive from a passive sensor.

### 12.2.1 An example of choosing between sensors

Flynn (1985) has used a simple set of rules to combine information from sonar and infra-red sensors. The sensors were mounted on a mobile robot Robart II and scanned around a room to determine the position of the walls, doorways, and assorted obstacles. Both of the sensors mounted on Robart II had very severe performance limitations.

#### *12.2.1.1 Sonar*

- Generates a large ultrasound beam width and therefore responds to quite small objects over a range of ±40° in bearing.
- Suffers from specular reflection — object surfaces appear mirror-like.
- Resolution is 0.12 in (3 mm) in the range 0.9–35 ft (0.27–10.7 m). However, sensor accuracy is worse than this even when compensated for air temperature and humidity.

#### *12.2.1.2 Infra-red*

- Receiver had a narrow angle of reception, therefore object bearing information is reasonably accurate.
- Range measurement is unreliable due to varying surface reflectance of items in the room. Maximum range is about 10 ft (3.05 m), depending on the surface finish.

Effectively, the sonar sensor provides an accurate measurement of the distance to an object, but poor resolution of its angular bearing. The characteristics of the infra-red detector are almost the reverse, with good estimation of bearing and poor measurement of range. The complementary characteristics of the sensors are illustrated in Figure 12.1.

#### *12.2.1.3 Rules for combining observations*

The following simple rules were used to combine the sonar and infra-red sensor data:

1. Whenever the infra-red sensor detects a change from detecting an object to no detection, and the associated sonar reading is less than 10 ft (3.05 m), then it is very likely that a valid depth discontinuity has been detected (like a doorway or isolated obstacle).

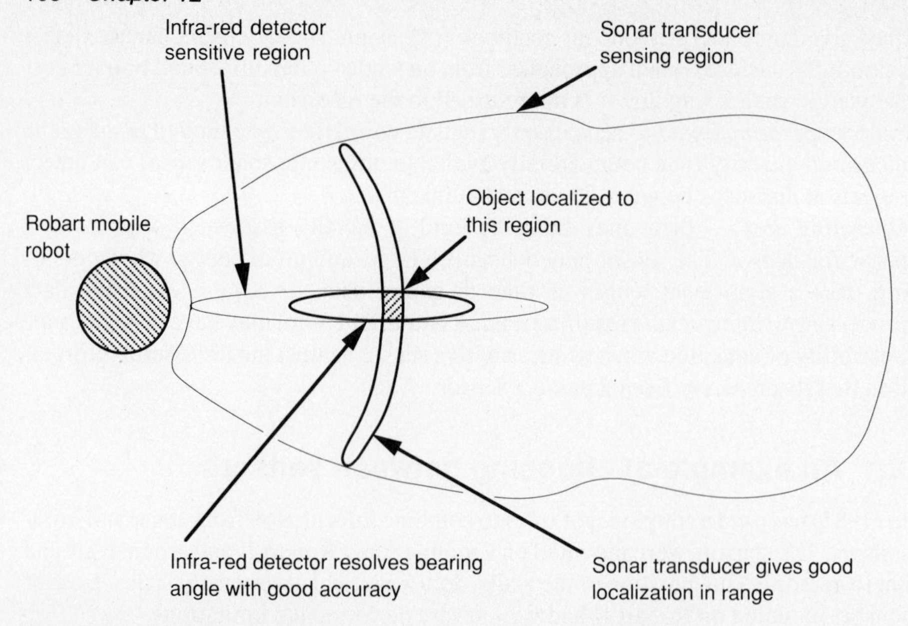


**Figure 12.1** Object localization using sonar and infra-red sensors
(*Note*: this diagram is not to scale.)

2. If the sonar reading is greater than the maximum range for the infra-red, then ignore the infra-red sensor.
3. If the sonar reading is at its maximum value, then the real distance is greater.

By applying these rules to sonar and infra-red sensor data, depth maps of rooms were created with greater accuracy than could be produced by one of the sensors alone. Each sensor was used to acquire the information (either range or bearing) that it could measure accurately.

## 12.3 Combining data from multiple sensors

The alternative to choosing a single sensor is the fusion of all sources of data into one model of the world using an analysis of the reliability of the information from each sensor. If a detailed knowledge of the error statistics for each sensor is available, several statistical methods have been suggested which can eliminate incorrect readings and merge the remainder into an accurate composite view.

Repeated measurements of the same quantity will show a random variation due to noise, quantization errors, and other disturbing effects. In very many cases, the spread of readings can be modeled as a normal or Gaussian distribution. This probability density function has the characteristic bell shape shown in Figure 12.2 and the equation:

$$p(x) = \frac{1}{\sqrt{2\pi\sigma^2}} e^{-\left(\frac{(x-\mu)^2}{2\sigma^2}\right)} \tag{12.1}$$

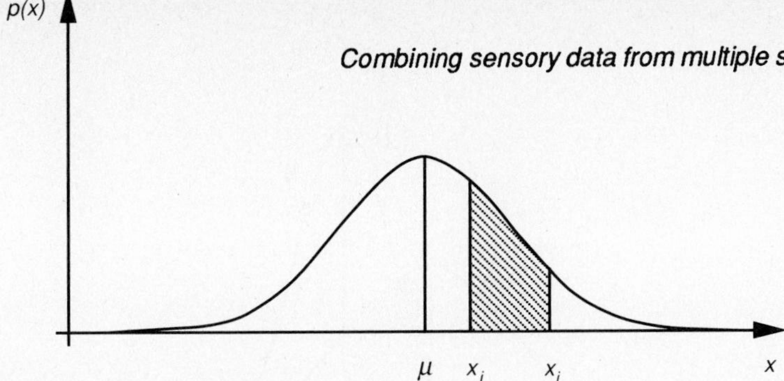


**Figure 12.2** The normal distribution

where:

$\mu$ = mean value of the distribution; and
$\sigma$ = the standard deviation.

If a sensor indicates a reading $\mu$ of quantity $x$ then the normal distribution can be used to determine the probability that the true value of $x$ lies between two specified values of $x$. The probability $p_{ij}$ that the true value of $x$ lies between $x_i$ and $x_j$ is given by:

$$p_{ij} = \frac{1}{\sqrt{2\pi\sigma^2}} \int_{x_i}^{x_j} e^{-\left(\frac{(x-\mu)^2}{2\sigma^2}\right)} dx \qquad (12.2)$$

This can be interpreted as the area under the normal curve between $x_i$ and $x_j$. The total area under the normal curve is 1. The normal distribution provides a compact means of classifying the reliability of sensor data in terms of the variance $\upsilon$ (square of the standard deviation $\sigma$) of the sensor output.

### 12.3.1 Eliminating erroneous sensor readings

The confidence distance has been proposed as the basis for a technique which allows spurious sensor readings to be eliminated in a multi-sensor environment (Luo et al. 1988). (The article by Luo et al. (1988) should be read in conjunction with the comments and corrections made by Mintz and Luo (1990).) Readings from two sensors A and B are shown in Figures 12.3 and 12.4 together with their associated probability density functions.

Confidence distance can be interpreted graphically as twice the area of one of the probability density functions between the limits $x_A$ and $x_B$. Because both A and B have a probability density function in this region, there are two confidence distances $d_{AB}$ and $d_{BA}$. Confidence distance $d_{AB}$ can be interpreted as the probability that $x_B$ is correct given the measured value of $x_A$. Thus $d_{AB}$ represents the amount of support that sensor reading $x_A$ gives to the reading $x_B$, and $d_{BA}$ the amount of support that sensor reading $x_B$ gives to the reading $x_A$. In practice, an empirical threshold is set for any pair of sensors X and Y, and if $d_{XY}$ is lower than that threshold then the data from sensor X 'agrees' with the data

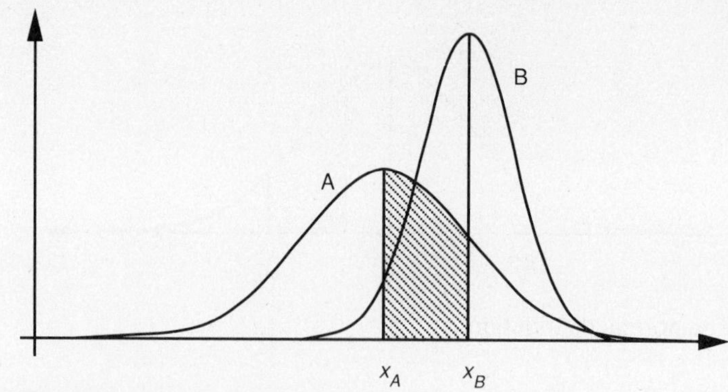


**Figure 12.3** Confidence distance measure $d_{AB}$
(Redrawn from Luo et al. 1988, ©1988 IEEE)

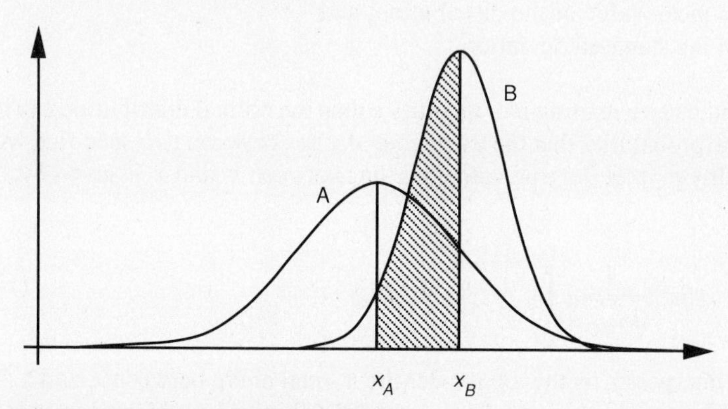


**Figure 12.4** Confidence distance measure $d_{BA}$
(Redrawn from Luo et al. 1988, ©1988 IEEE)

from sensor Y. The data from multiple sensors is examined to find a group of readings that are all in mutual agreement with each other. Any sensor reading not in this group is then treated as unreliable.

## 12.3.2 Fusing data from multiple sensors

The variance can be used to fuse multiple sensor readings into one composite value. An ad hoc technique weights the sensor readings inversely with the sensor variance. The combined sensor reading $\xi$ resulting from the output $x_A$ of sensor A (standard deviation $\sigma_A$) and output $x_B$ of sensor B (standard deviation $\sigma_B$) is given by the following equation:

$$\xi = x_A + (x_B - x_A)\frac{\sigma_A^2}{(\sigma_A^2 + \sigma_B^2)} \tag{12.3}$$

Figures 12.5(a), 12.5(b), and 12.5(c) show the result of combining data from two sensors where one of the sensors has different values of variance. As expected, the combined value $\xi$ lies halfway between the two sensor measurements when the sensors have equal variance and tends towards the more reliable sensor (lower variance) when the sensors have differing variances.

Many other techniques have been developed for data fusion, including some based on Bayes Theorem (Durrant-Whyte 1987) and Kalman filtering (Moutarlier and Chatila 1989). A major problem in sensor fusion is to determine the correspondence between data from different sensors. This problem occurs in stereo vision and amounts to finding regions in the images from both the left and right eyes that correspond to the same physical object in the scene being viewed.

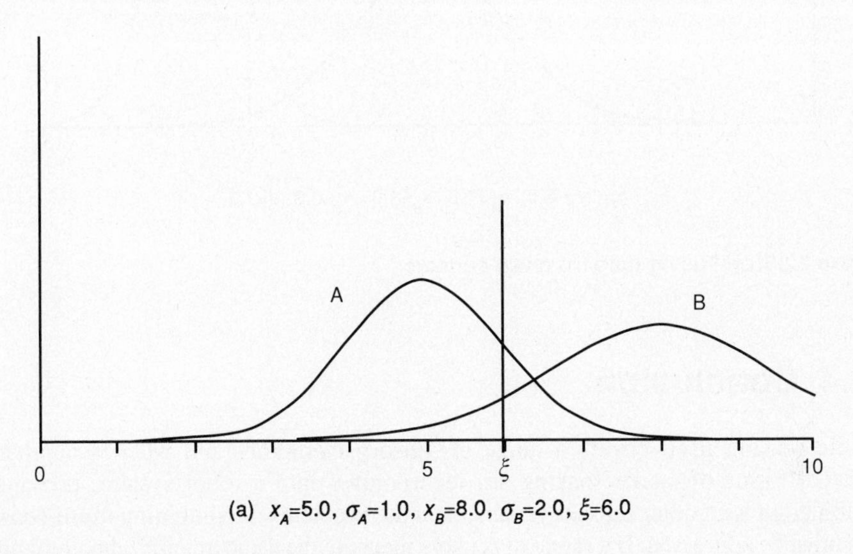


**Figure 12.5(a)** Fusing data from two sensors

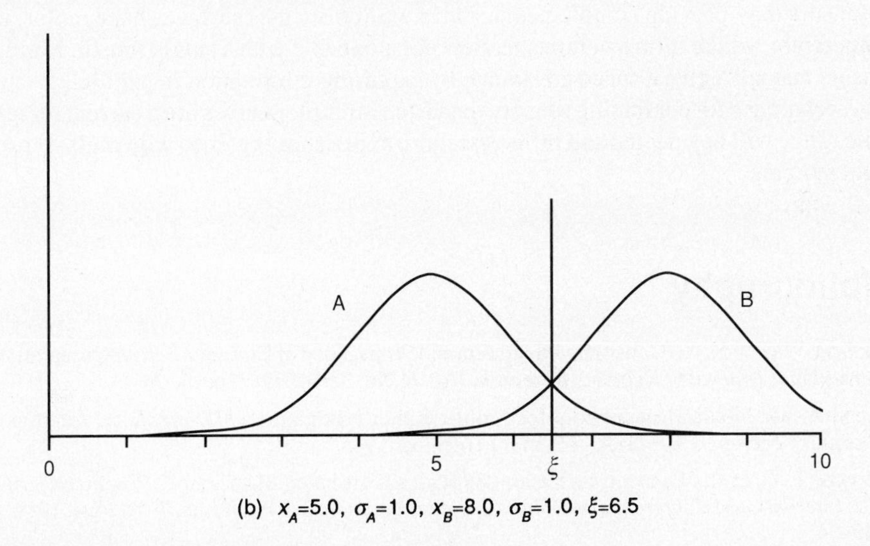


**Figure 12.5(b)** Fusing data from two sensors

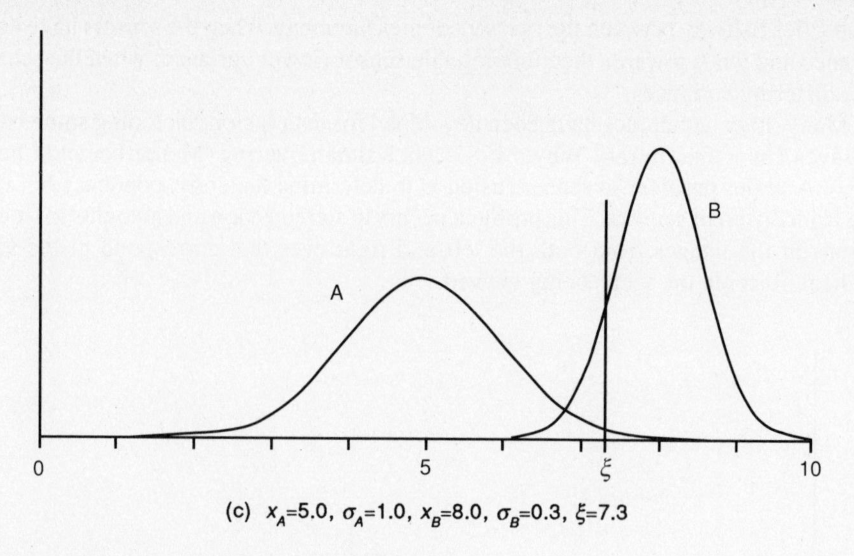


**Figure 12.5(c)** Fusing data from two sensors

## 12.4 Conclusion

Tactile sensing itself covers a range of sensory modalities and when a number of different forms of tactile sensing are incorporated into a robot system, perhaps in combination with other sensor systems, then the problems of combining multi-sensory data must be addressed. If a range of sensors measure the same quantity then redundant information can be analyzed to give more accurate information. Alternatively, a group of sensors may provide complementary information such as surface shape, color, and temperature, which form a composite view not attainable with a single sensor. Multiple sensors may also give a speed advantage by capturing information in parallel.

Techniques for combining sensory data from multiple sources are a current research topic. They will be selected and refined as more experience is gained with multi-sensory robot systems.

## Bibliography

DURRANT-WHYTE, H. F., 'Consistent Integration and Propagation of Disparate Sensor Observations', *International Journal of Robotics Research*, Vol. 6, No. 3, Fall 1987, pp. 3–24.

FLYNN, A. M., 'Redundant Sensors for Mobile Robot Navigation', *MIT Artificial Intelligence Laboratory Technical Report AI-TR-859*, September 1985.

HARMON, S. Y., et al., 'Sensor Data Fusion Through a Distributed Blackboard', *Proceedings of the IEEE International Conference on Robotics and Automation*, San Francisco, 7–10 April 1986, pp. 1449–54.

LUO, R. C., et al., 'Dynamic Multi-Sensor Data Fusion System for Intelligent Robots', *IEEE Journal of Robotics and Automation*, Vol. 4, No. 4, August 1988, pp. 386–96.

LUO, R. C., and KAY, M. G., 'Multisensor Integration and Fusion in Intelligent Systems', *IEEE Transactions on Systems, Man and Cybernetics*, Vol. 19, No. 5, September/October 1989, pp. 901–31.

MINTZ, M., and LUO, R. C., 'Comments on "Dynamic Multi-Sensor Data Fusion System for Intelligent Robots" ', *IEEE Transactions on Robotics and Automation*, Vol. 6, No. 1, February 1990, pp. 104–6.

MOUTARLIER, P., and CHATILA, R., 'Stochastic Multisensory Data Fusion for Mobile Robot Location and Environment Modelling', *Proceedings of 5th International Symposium of Robotics Research*, 28–31 August 1989.

## Questions

12.1 When a number of sensors are available to measure the same quantity, explain the criteria which can be used to select the single 'best' sensor to make the measurement.

12.2 Describe a sensing application where multiple sensors were available and sensor data was combined by choosing the 'best' sensor to provide each desired quantity.

12.3 Outline a technique for determining the reliability of sensor data based on confidence distance.

12.4 Describe a technique for fusing the data from multiple sensors based on weighting the data with the inverse variance of the sensor output.

12.5 Many sensors fail to function correctly under some combinations of operating conditions. Conduct a survey of camera autofocus techniques (ultrasonic, infrared, etc.) and determine under what conditions they give incorrect results even though operating within their specified range limits. Determine the physical reason for the failures.

# Index